Dark Energy

Peking University–World Scientific Advance Physics Series

ISSN: 2382-5960

Series Editors: Enge Wang *(Peking University, China)*
Jian-Bai Xia *(Chinese Academy of Sciences, China)*

Peking University-World Scientific Advance Physics Series

Vol 1

Dark Energy

Miao Li • Xiao-Dong Li • Shuang Wang
Chinese Academy of Sciences, China

Yi Wang
McGill University, Canada

W͞ World Scientific

NEW JERSEY • LONDON • SINGAPORE • BEIJING • SHANGHAI • HONG KONG • TAIPEI • CHENNAI

Published by

World Scientific Publishing Co. Pte. Ltd.

5 Toh Tuck Link, Singapore 596224

USA office: 27 Warren Street, Suite 401-402, Hackensack, NJ 07601

UK office: 57 Shelton Street, Covent Garden, London WC2H 9HE

Library of Congress Cataloging-in-Publication Data
Li, Miao, 1962–   author.
   Dark energy / Miao Li (Chinese Academy of Sciences, China), Xiao-Dong Li (Chinese Academy
of Sciences, China), Shuang Wang (Chinese Academy of Sciences, China), Yi Wang (McGill
University, Canada).
      pages cm -- (Peking University-World Scientific advance physics series ; volume 1)
   Includes bibliographical references and index.
   ISBN 978-9814619707 (hardcover : alk. paper)
   1. Dark energy (Astronomy) I. Li, Xiao-Dong (Researcher on physics), author. II. Wang, Shuang
(Researcher on physics), author. III. Wang, Yi (Researcher on physics), author. IV. Title.
   QB791.3.L53 2014
   523.01--dc23

                                                             2014022440

British Library Cataloguing-in-Publication Data
A catalogue record for this book is available from the British Library.

Cover image credit: NASA, ESA, E. Jullo (JPL/LAM), P. Natarajan (Yale) and J.-P. Kneib (LAM).

The work is originally published by Peking University Press in 2012.

In-house Editor: Ng Kah Fee

Foreword

Physics is the study of matter and energy, and their mutual interactions. Physics research provides an elementary driving force for mankind's intellectual development. It lays a solid foundation for a wide range of scientific disciplines, including chemistry, life science, materials science, information technology, energy and environmental sciences. It also leads many cutting-edge interdisciplinary fields of research. In the modern world of rapid technological innovation, physics has transcended its traditional realm as a branch of fundamental and applied science, and is playing an ever more important role in societal progress and in improving the general quality of life.

In the past three decades or so, the steady development of China's socio-economical system has enkindled unprecedented cultural and intellectual advancements. At the same time, Chinese physics community has made tremendous progresses, and is rising rapidly in the international arena. Chinese physicists are experiencing a golden age unparalleled in the history of China. Concomitantly, the publishing of scholarly books and monographs in physics has become a flourishing industry, which will inevitably play an indispensible role in the transmission of knowledge, scholarly interactions and the education of specialists as well as the general public.

It is in this context that Peking University (Press) publishes this "Peking University-World Scientific Advance Physics Series", in a collaborative effort with the World Scientific Publishing Company, the largest English-language academic publisher in the Asia-Pacific region. The Series aims to present the most significant achievements of Chinese physicists, both domestic and international, and to introduce some of their best books and monographs to the international community. Published in English, this Series will share with international readers the history and present-day status of Chinese physics community.

I am very pleased to witness the publishing of the "Peking University-World Scientific Advance Physics Series", a joint effort of Peking University (Press) and World Scientific. I believe this Series will bring forth novel insights, intellectual challenges and simple joy of reading to those who love and study physics, and also offers unique perspectives to other academic disciplines, as a contribution of the Chinese physics community to the global scientific and technological development.

Enge Wang
President, Peking University
Professor of Physics

Preface

The dark energy problem is a longstanding issue:

In 1917, Einstein added a cosmological constant term in his field equation, to obtain a static Universe. However, he removed this cosmological constant in 1931 because of the discovery of the cosmic expansion. Due to losing the opportunity to predict the expansion of the Universe, Einstein claimed that this is the biggest blunder in his life.

In 1967, Zel'dovich reintroduced the cosmological constant by taking the vacuum fluctuations into account. But in quantum field theory, the expectation value of the vacuum fluctuations diverges as k^4. If we take a cutoff at the Planck or electroweak scale, the predicted value of the cosmological constant Λ is still 117 or 52 orders of magnitude larger than the observed value. This discrepancy is known as the "fine-tuning problem", and has perplexed theorists for several decades.

In 1998, Based on the analysis of 34 nearby and 16 distant type Ia supernovae, Riess *et al.* first discovered the acceleration of expansion of the Universe. Soon after, utilizing 18 low-redshift and 42 high-redshift supernovae, Perlmutter *et al.* confirmed the discovery of cosmic acceleration. This discovery declares the return of the cosmological constant, now termed as "dark energy". This leads to the so-called "coincidence problem": why the current dark energy density is of the same order of matter density?

To solve these puzzles and understand the cosmic acceleration, in the last decade, theorists have proposed numerous theoretical models and have written thousands of papers on this subject. Unfortunately, so far the nature of dark energy is still a mystery. On the observational side, astronomers have also spent enormous efforts on probing dark energy. Since 1998, observational evidences of dark energy have been definitely confirmed and highly improved through many cosmological observations, such as the type Ia supernovae, the baryon acoustic oscillations, the cosmic microwave background, the weak gravitational lensing, the galaxy cluster survey, and so on. Therefore, the dark energy problem, which challenges our understanding of fundamental physical laws and the nature of the cosmos, has become one of the central problems in theoretical physics and modern cosmology. In this situation, a book introducing the current state of research on dark energy is very necessary. We write this book based on those considerations. The book is an extended version of our previous review article [Commun. Theor. Phys. **56** (2011) 525], and it consists of three parts:

In the first part, we start with an introduction to preliminary knowledge. Fundamentals for a modern understanding of cosmology, including general relativity, matter components and FRW cosmology, are summarized and briefly introduced. Thus this part serves as a preparation and foundation for our introduction to the dark energy in the next two parts.

The second part reviews some of major theoretical ideas and models of dark energy. In this part, we first briefly review the history of the problem of the cosmological constant. After this, we follow Weinberg's classical review in 1989, and divide the old approaches reviewed by Weinberg into five categories. Then, we add three more categories in order to include most of the more recent ideas and models. Eight classes of dark energy models are reviewed: models based on symmetry; anthropic principle; tuning mechanism; modified gravity; quantum cosmology; holographic principle; back-reaction and phenomenological constructions.

The third part is devoted to reviewing some observational and numerical works. First, we will introduce some basic knowledge of statistics. Then, we will review the mainstream cosmological observations probing dark energy. Next, we will provide a brief overview of the present and future dark energy projects. Finally, we will review the current numerical studies on cosmic acceleration, including the observational constraints on specific theoretical models, and the model-independent reconstructions from observational data.

The aim of this book is to provide a sufficient level of understanding of dark energy problem, so that the reader can both get familiar with this area quickly and also be prepared to tackle the scientific literature on this subject. It will be useful for graduates students and advanced undergrads in physics or astronomy to obtain an overall glance of the current dark energy research, and it will also be helpful for researchers who are interested in this subject to design their own research programs.

We would like to express our appreciation for the kind invitation of the Peking University Press. We are also very grateful to our editor Xiao Liu, for the kind support and encouragement. Many thanks to Prof. Xin Zhang for valuable suggestions about the contents and styles of this book. We thank Prof. Robert Brandenberger, Dr. Yi-Fu Cai, Prof. Qing-Guo Huang and Prof. Richard Woodard for the useful discussions. We also thank Prof. Robert Kirshner for the helpful suggestions.

Miao Li, Xiao-Dong Li, Shuang Wang
Institute of Theoretical Physics,
Chinese Academy of Sciences,
Beijing 100190, China
Department of Modern Physics,
University of Science and Technology of China,
Hefei 230026, China

Yi Wang
Physics Department, McGill University,
Montreal, H3A2T8, Canada

Contents

Part I

Preliminaries in a Nutshell

1
Gravitation

For a quick review of gravitation, we start from the equations of Newtonian gravity. The gravitational force acting on a test mass m is

$$F = -m\nabla\Phi\,, \tag{1.1}$$

where Φ is called the Newtonian potential, which is determined by the Poisson equation

$$\Delta\Phi = 4\pi G\rho\,, \tag{1.2}$$

where ρ describes the mass distribution of a gravitational source, and G is the Newton's gravitational constant.

Newtonian gravity captures some features of our universe. However, mass sums up. As a result, on cosmological scales, Newtonian gravity breaks down. For a systematic description of our universe, the theory of general relativity is inevitably required.

In the remainder of this chapter we briefly review the two key concepts in general relativity: what if spacetime is curved and what makes spacetime curved. Here we are by no means trying to provide a logical introduction for general relativity. Instead, we pragmatically highlight parts of the theory which are building blocks for a phenomenological description of dark energy, as well as set up our conventions.

1.1   The Curved Spacetime

It is for a long time a great mystery why our space is well described by Euclidean geometry. In the spatial two dimensional story, numerous attempts are made to prove the "parallel axiom" and eventually the axiom turns out not provable. Instead, Euclidean geometry only applies for flat slicings, and the parallel axiom is the additional assumption to be made.

In the 3-dimensional space story, similarly the fact that our spatial dimensions are well described by Euclidean geometry is not for granted but instead has profound implications in physics, which in part leads to the discovery of cosmic inflation. Moreover, in the 3+1 dimensional story, when time is also considered, the geometry of spacetime is proven not flat on cosmological spacetime scales. Thus a description of curved spacetime is an essential in modern cosmology.

For two dimensional curved space, an intuitive understanding can be obtained by embedding into 3 dimensions. However, to investigate our curved spacetime, it is more convenient to use an intrinsic description instead of the embedding way. This is because on the one hand, we do not have good intuition on higher dimensions where our spacetime can be embedded into, and on the other hand, although such embeddings always exist in mathematics, there is no evidence for their physical existence.

The metric is what one needs to construct an intrinsic description of spacetime. A proper length between two nearby points can be defined as

$$ds^2 = g_{\mu\nu}(x)dx^\mu dx^\nu \,, \tag{1.3}$$

where $g_{\mu\nu}(x)$ is the metric, which describes the geometry of spacetime. We require the metric to be symmetric: $g_{\mu\nu} = g_{\nu\mu}$, because the anti-symmetric part does not contribute as long as dx^μ and dx^ν commute.

Our spacetime is Lorentzian. In other words, there is one time dimension and all the other dimensions are spatial. The metric takes different sign between the two. The sign convention we use here is $(-,+,+,+)$. Greek labels as μ and ν take value in these dimensions, and repeated labels are by default summed (Einstein's convention).

The affine connection (also known as the Christoffel symbol), describing how a vector moves along a curve, is defined as

$$\Gamma^\lambda{}_{\mu\nu} = \frac{1}{2}g^{\lambda\rho}\left(\partial_\mu g_{\rho\nu} + \partial_\nu g_{\rho\mu} - \partial_\rho g_{\mu\nu}\right), \tag{1.4}$$

where ∂_μ is a shorthand for partial derivative $\partial/\partial x_\mu$. The affine connection is symmetric about the two lower indices: $\Gamma^\lambda{}_{\mu\nu} = \Gamma^\lambda{}_{\nu\mu}$.

Covariant derivatives can be constructed using the connection as

$$\nabla_\mu \phi = \partial_\mu \phi \,, \tag{1.5}$$

$$\nabla_\mu A_\nu = \partial_\mu A_\nu + \Gamma^\lambda{}_{\mu\nu} A_\lambda \,, \qquad (1.6)$$

$$\nabla_\mu A^\nu = \partial_\mu A^\nu - \Gamma^\nu{}_{\mu\lambda} A^\lambda \,, \qquad (1.7)$$

$$\cdots$$

where ϕ is a scalar field, and A_ν and A^ν are vector fields with lower and upper indices respectively. Covariant derivatives of higher rank tensors can be constructed similarly.

The Riemann curvature tensors can be constructed based on the metric, the affine connection and their derivatives. Among those the most useful ones are the Riemann tensor $R^\rho{}_{\sigma\mu\nu}$, the Ricci tensor $R_{\mu\nu}$, and the Ricci scalar $\mathcal{R}$, defined as

$$R^\rho{}_{\sigma\mu\nu} = \partial_\mu \Gamma^\rho{}_{\nu\sigma} - \partial_\nu \Gamma^\rho{}_{\mu\sigma} + \Gamma^\rho{}_{\mu\lambda}\Gamma^\lambda{}_{\nu\sigma} - \Gamma^\rho{}_{\nu\lambda}\Gamma^\lambda{}_{\mu\sigma} \,, \qquad (1.8)$$

$$R_{\mu\nu} = R^\lambda{}_{\mu\lambda\nu} \,, \qquad \mathcal{R} = g^{\mu\nu} R_{\mu\nu} \,. \qquad (1.9)$$

Note that $R_{\mu\nu} = R_{\nu\mu}$ is symmetric.

A free particle (which is not affected by any other force except gravitation) moves following a line that extremizes its proper distance. Such a line is called a geodesic. Supposing the free particle moves along a world line $x^\mu(\lambda)$, the world line satisfies the geodesic equation

$$\frac{d^2 x^\rho}{d\lambda^2} + \Gamma^\rho{}_{\mu\nu} \frac{dx^\mu}{d\lambda} \frac{dx^\nu}{d\lambda} = 0 \,. \qquad (1.10)$$

Defining four velocity $u^\rho \equiv dx^\rho/d\lambda$, the geodesic equation can also be written as

$$\frac{du^\rho}{d\lambda} + \Gamma^\rho{}_{\mu\nu} u^\mu u^\nu = 0 \,. \qquad (1.11)$$

It is in principle arbitrary to choose the parameterization λ. However for massive particles a particular convenient choice is $d\lambda = d\tau$, where τ is the proper time. This choice of λ is called affine parameter and in this case the four velocity is normalized to $u^\mu u_\mu = -1$.

1.2   The Curved Spacetime: An Example

Here we illustrate the concepts of the last section by considering an explicit (and useful) example. Consider the metric called "spatially flat Friedmann–Robertson–Walker metric":

$$ds^2 = -c^2 dt^2 + a(t)^2 dx^i dx^i \,, \qquad (1.12)$$

where c is the speed of light. Hereafter, unless otherwise noted, we will take $c = 1$. The nonzero components of the connection can be worked out as

$$\Gamma^1{}_{01} = \Gamma^2{}_{02} = \Gamma^3{}_{03} = H\,, \qquad \Gamma^0{}_{11} = \Gamma^0{}_{22} = \Gamma^0{}_{33} = a^2 H\,, \qquad (1.13)$$

where $H \equiv \dot{a}/a$ is the Hubble parameter. The dot on a variable denotes time derivative. The value of H at the present epoch (denoted as H_0) is called the Hubble constant.

The components of the Ricci tensor are

$$R_{11} = R_{22} = R_{33} = 3a^2 H^2 + a^2 \dot{H}\,, \qquad R_{00} = -3a^2 H^2 - 3a^2 \dot{H}\,, \qquad (1.14)$$

and the Ricci scalar takes the form

$$\mathcal{R} = 6\dot{H} + 12H^2\,. \qquad (1.15)$$

The kinematics of a massless particle can be derived from the geodesic equation. Or alternatively, here we use a simpler method: a massless particle follows a null geodesic. In other words, $ds = 0$. In this sense, the metric can be thought of as an equation of motion:

$$dr = \frac{dt}{a(t)}\,, \qquad R_{\mathrm{h}} = a(t) \int_{t_{\mathrm{i}}}^{t_{\mathrm{f}}} \frac{dt'}{a(t')}\,, \qquad (1.16)$$

where $R_{\mathrm{h}} \equiv a(t)r$ is the physical distance that the massless particle has traveled from t_{i} to t_{f}. The distance is measured at time t. The equation (1.16) looks simple. However there is interesting and deep physics hiding behind.

For some forms of $a(t)$, R_{h} may keep finite when taking $t_{\mathrm{i}} \to -\infty$. In this case, light travels only a finite distance from the infinite past to now (the time t_{f}). In other words, not all particles in the universe can reach us at the present time. Thus in this case R_{h} is called a particle horizon.

For some forms of $a(t)$, R_{h} may keep finite when taking $t_{\mathrm{f}} \to \infty$. In this case, light could only travel a finite distance from now to the infinite future. In other words, even we live long enough and keep making observations, there exist spatial regions in our universe that we can never observe. Thus in this case R_{h} is called an event horizon, or more specifically future event horizon.

Note that we have proposed an ansatz (1.12) in the discussion. Thus the situations of horizons include, but not limited to the cases discussed here.

Let us refocus our attention to the motion of a massless particle. Apart from the geodesics, the energy of the particle also changes while the particle is propagating. As a result of adiabaticity of the cosmic expansion,[1] the wavelength of the massless particle has to be proportional to $a(t)$. Thus the energy is redshifted as $E \propto 1/a(t)$.

The case of a massive particle is also interesting. In some sense, a massless particle can be thought of as a limit of a massive particle. Thus a massive particle should feel some analog of redshift effect as well. Indeed the wave packet of a massive particle is stretched by the dynamics of the universe thus the momentum of the particle decreases. Another way to see the same effect is the equivalence principle. In an inertial frame, the curvature of space behaves as gravitational force. Massless particles are pulled by the gravitational force, massive particles should also get pulled.

To see this explicitly, consider a free massive particle, with an initial velocity v, moving in flat FRW space time. The geodesic equation takes the form

$$\frac{du^i}{d\lambda} + Hu^0 u^0 = 0, \qquad \frac{du^0}{d\lambda} + a^2 Hu^i u^i = 0. \tag{1.17}$$

At first sight, the above equations may appear difficult because they are non-linear and coupled. However, the solution turns out not difficult at all. Note that $u^\mu u_\mu = -1$. Thus defining $u \equiv \sqrt{g_{ij} u^i u^j}$, we have $u^0 du^0 = u du$. Also, note $u^0 = dt/d\lambda$ (the 0-th component of $u^\mu = dx^\mu/d\lambda$). Then one obtains

$$\frac{du}{dt} + Hu = 0, \qquad u \propto 1/a. \tag{1.18}$$

Thus a freely moving massive particle will feel a friction force in an expanding background, and will eventually get comoving at late times, if the particle does not feel any other forces.

1.3   The Einstein Equation

The dramatic fact of general relativity is that spacetime is not only curved, but also dynamical. In other words, not only matter motion is affected

---

[1] Here assuming the wavelength of the particle is much shorter than the Hubble radius. In this case the expansion of the universe is adiabatic in the thermodynamical sense. Thus the expansion of the universe does not lead to entropy increase or particle production.

by curved space, but also matter creates curvature in space. How matter sources gravity is described by the Einstein equation

$$G_{\mu\nu} \equiv R_{\mu\nu} - \frac{1}{2}g_{\mu\nu}\mathcal{R} = 8\pi G T_{\mu\nu}\,, \qquad (1.19)$$

where $T_{\mu\nu}$ is the stress tensor, which will be reviewed in the next chapter. One could also add a cosmological constant term $-\Lambda g_{\mu\nu}$ to the right hand side of Eq. (1.19). We shall consider this possibility when reviewing the stress tensor.

As can be easily checked, the Einstein tensor $G_{\mu\nu}$ is covariantly conserved: $\nabla_\mu G^{\mu\nu} = 0$.

The Einstein equation (1.19) tells that some of the components in the metric have second time derivatives in its equation of motion, and thus propagating. Other components in the metric have at most one time derivative, and thus determined by matter distribution on the right hand side of the equation.

It is also useful to write the Einstein equation in terms of an action principle. On the one hand, turns out the equation takes a more elegant form. On the other hand, an action principle is a better starting point for some theoretical studies such as quantization. The Einstein equation can be derived from varying the following Einstein–Hilbert action

$$S = S_{\text{gravity}} + S_{\text{matter}}\,, \qquad S_{\text{gravity}} \equiv \frac{M_{\text{P}}^2}{2} \int d^4x \sqrt{-g}\mathcal{R}\,, \qquad (1.20)$$

where $\sqrt{-g}$ is a shorthand of $\sqrt{-\det(g_{\mu\nu})}$, and $M_{\text{P}} \equiv 1/\sqrt{8\pi G}$ is called the reduced Planck mass . Sometimes we also call M_{P} Planck mass for short. The gravitational action S_{gravity} is called the Einstein–Hilbert action.

From the action Eq. (1.20), we can derive the Einstein equation either by varying the metric only or also the connection. The former is most frequently used and is called the metric formulation of general relativity. The variation of the gravitational part and matter part are proportional to the left and right hand side of Eq. (1.19), respectively:

$$G_{\mu\nu} = \frac{2}{M_{\text{P}}^2\sqrt{-g}}\frac{\delta S_{\text{gravity}}}{\delta g^{\mu\nu}}\,, \qquad (1.21)$$

$$T_{\mu\nu} = -\frac{2}{\sqrt{-g}}\frac{\delta S_{\text{matter}}}{\delta g^{\mu\nu}}\,, \qquad (1.22)$$

where to calculate $\sqrt{-g}$, we have used $\delta g = gg^{\mu\nu}\delta g_{\mu\nu} = -gg_{\mu\nu}\delta g^{\mu\nu}$. Also, the $\sqrt{-g}g^{\mu\nu}\delta R_{\mu\nu}$ term is a total derivative thus does not contribute to the variation principle, thanks to the simplicity of the Einstein–Hilbert action.

2
Matter Components

2.1 The Stress Tensor

The stress tensor (alternatively called the energy-momentum tensor) describes the amount of energy, momentum, and their fluxes in a unit volume.

The tensor is named "stress" because in Newtonian mechanics (and not necessarily fluid mechanics), the stress tensor (as a 3×3 matrix) is a measure of stress. A matrix instead of a vector is needed because, consider a cross section of some material, the force acting on the cross section may not be perpendicular to the cross section. There may be also parallel components (shear force). According to Cauchy, such a case should be described by a 3×3 matrix: the stress tensor.[1]

The ij component of a stress tensor denotes what is the i component of a force acting on a cross section (with unit mass), whose normal points to the j direction. If the material is in equilibrium, the summation of moments acting on any volume element has to be zero. As a result, the stress tensor has to be symmetric for equilibrium material.

It is straightforward to generalize the stress tensor to relativity. Now both space and force becomes 4-dimensional thus the stress tensor in a coordinate system becomes a 4×4 matrix. Also, as the stress tensor in general relativity is defined as variation of the matter action with respect to the metric by Eq. (1.23), the stress tensor in general relativity is automatically symmetric in its indices.

In classical general relativity, the stress tensor is the unique input to the gravitational sector telling information about the matter sector.

---

[1] The stress tensor is a generalization of pressure, which has only one component instead of 6. As we shall see in cosmology, for most of our discussion, the stress tensor is actually reduced into pressure. However, as a development of general formalism, such a generalization $\longrightarrow$ reduction procedure is worthy.

The stress tensor is locally conserved: $\nabla_\mu T^{\mu\nu} = 0$. The local conservation of the stress tensor can be understood in the following two aspects.

On the one hand, $\nabla_\mu T^{\mu\nu} = 0$ is a direct generalization of $\partial_\mu T^{\mu\nu} = 0$. In Newtonian mechanics or special relativity, $\partial_\mu T^{\mu\nu} = 0$ indicates that energy (considering both matter and gravity) is globally conserved, because in flat space one can always integrate the equation. However, in general relativity, $T^{\mu\nu}$ only denotes matter energy. Also space is only known to be flat in infinitesimal volume elements, thus the conservation of matter energy holds only locally. To have global concept of energy, one need to define gravitational energy with care. For example, if one defines gravitational energy from the Hamiltonian constraint, then the gravitational energy of a system is always the negative of matter energy, and the total energy sums to zero.

On the other hand, for any matter component satisfying Einstein equation, the local conservation of stress tensor directly follows from $\nabla_\mu G^{\mu\nu} = 0$. In other words $\nabla_\mu T^{\mu\nu} = 0$ can be derived from Einstein equations. Thus for a fluid with a single component, the equation of motion for the fluid can be derived from the Einstein equations, up to some integration constants. For fluids with multiple components, $\nabla_\mu T^{\mu\nu} = 0$ will be a consistency relation between the gravitational sector and the matter sector.

2.2 Perfect Fluid

When a matter component can be considered as continuous, the component can be considered by continuum mechanics. In addition, if the component cannot support shear stress,[2] the fluid description applies. The fluid mechanics description is particularly useful, because it provides a unified framework for considering most of the usual matter components in cosmology.

In these introductory chapters, we will focus on perfect fluid, that viscosity and heat conduction can be neglected.

First let us consider an infinitesimal volume element of the fluid, and choose a comoving frame with respect to the volume, such that in the coordinate system the volume in the fluid has no velocity. In this case $T^0{}_i = 0$ automatically. The non-diagonal components of $T^i{}_j$ also vanishes

---

[2] For non-perfect fluid, there can be shear stress (related to viscosity). Here "cannot support" means the shear stress in the fluid is not large enough to hold the shape of the fluid in equilibrium. This is a technical way to say that the fluid takes the shape of its container.

as a result of the lack of viscosity. Finally, as the fluid does not have over all moving direction, the pressure in the three directions should be the same. To conclude, the perfect fluid element in a rest frame has stress tensor[3]

$$T^{\mu}{}_{\nu} = \mathrm{diag}(-\rho, p, p, p). \tag{2.1}$$

One can then perform a coordinate transformation to write the stress tensor in an arbitrary coordinate, where the fluid element under consideration has a velocity $u^{\mu}(x)$:

$$T_{\mu\nu} = (\rho + p)u_{\mu}u_{\nu} + g_{\mu\nu}p. \tag{2.2}$$

2.3 Observers and Energy Conditions

It is useful to relate the stress tensor with the energy-momentum density measured by an observer. Consider an observer moving with velocity u^{μ}, the energy-momentum density measured is $T_{\mu\nu}u^{\mu}$. Further, the energy density measured is $T_{\mu\nu}u^{\mu}u^{\nu}$.

We shall discuss the stress tensor for different matter components in the reminder of this chapter. However, before that, it is useful to review some general considerations: What kind of stress tensor are considered to be "reasonable", at various levels. Energy conditions are useful for this purpose.

The null energy condition asserts that

$$T_{\mu\nu}u^{\mu}u^{\nu} \geqslant 0, \tag{2.3}$$

for every null vector field u^{μ}. In the ideal fluid description, this is to mean $\rho + p \geqslant 0$. The null energy condition is important in cosmology. As we shall see, it is the divide of whether the universe undergoes inflation or super-inflation; or whether the universe will develop a singularity or a bounce solution. On the other hand, the null energy condition is also a border line beyond which one need to be extremely careful because a number of physical laws easily got broken.

The null energy condition is very weak (actually weaker than the weak energy condition, as we shall introduce). However, one can still integrate

---

[3] There we require $T^{\mu}{}_{\nu}$, instead of $T_{\mu\nu}$ or $T^{\mu\nu}$ to be $\mathrm{diag}(-\rho, p, p, p)$. This is because although general relativity locally reduces to special relativity, the metric does not necessarily reduce to $\mathrm{diag}(-1, 1, 1, 1)$, but rather may up to a coordinate transformation. The up-down indices tensor will help us out from this situation, because "$g^{\mu}{}_{\nu}$" is always $\delta^{\mu}{}_{\nu}$.

the null energy condition along a flow line or spatial surface to obtain even weaker energy conditions.

The weak energy condition asserts that

$$T_{\mu\nu}u^\mu u^\nu \geqslant 0 , \tag{2.4}$$

for every time like u^μ (thus for every observer). In the ideal fluid description, this is to mean $\rho + p \geqslant 0$, and at the same time $\rho \geqslant 0$. The latter condition indicates that the weak energy condition, despite its name, is perhaps too strong. Because it rules out one of the most interesting geometries in string theory: the AdS spacetime, which is believed to be one of the best understood examples of quantum gravity. However, in cosmology, in most well known scenarios, the energy density is positive. Thus the weak energy condition does not differ too much from the null energy condition in those cosmological considerations.

The week energy condition can be generalized to the dominate energy condition, which asserts that the week energy condition holds, and at the same time $T_{\mu\nu}v^\mu$ is either time like or null. This additional restriction indicates $\rho^2 \geqslant p^2$ for an ideal fluid.

The strong energy condition:

$$\left(T_{\mu\nu} - \frac{1}{2}g_{\mu\nu}T \right) u^\mu u^\nu \geqslant 0 , \tag{2.5}$$

for every time like u^μ. In the ideal fluid language, this is to mean $\rho + p \geqslant 0$ and at the same time $\rho + 3p \geqslant 0$. The strong energy condition is actually a pure geometrical requirement. Because from Einstein's equation, the strong energy condition can be rewritten as $R_{\mu\nu}u^\mu u^\nu \geqslant 0$.

The strong energy condition, which appears more complicated in its form, is essential for proving the focusing theorem (that gravity is attractive). Thus the strong energy condition is too strong to allow any acceleration of the universe. On the other hand, we need a period of acceleration in our universe to explain the large scale correlations on the CMB. The observed dark energy also contradicts the strong energy condition. Yet another example is that domain walls break this energy condition. Thus the strong energy condition is perhaps also too strong a statement.

There is also a trace energy condition requiring $T^\mu{}_\mu \leqslant 0$. This is to mean $\rho - 3p \geqslant 0$ for an ideal fluid. This condition is known to contradict the equation of state (defined as $w \equiv p/\rho$) for neutron stars. Thus this condition is now abandoned.

2.4　The Vacuum

What do we mean when saying our world is Lorentz symmetric? We are not saying a car looks the same when it is moving towards or away from us. Instead, we are saying when there is "nothing" in the system, an inertial observer will never find whether the system is boosted or not. In other words, the ground state of the system, or the vacuum, is Lorentz invariant.

Now what is the stress tensor of this vacuum state? The form of the vacuum stress tensor is highly restricted by symmetries: namely $T_{\mu\nu}$ symmetric on permuting μ and ν, and the Lorentz symmetry.

It is straightforward to verify the following fact: a 2-dimensional tensor that is invariant under 2-dimensional rotation, must be of the form

$$\begin{pmatrix} x & y \\ -y & x \end{pmatrix},$$

thus a symmetric 2-dimensional tensor invariant under 2-dimensional rotation must be of the form $\mathrm{diag}(x, x)$. Similarly, if a 2-dimensional symmetric tensor is invariant under 2-dimensional boost, the tensor must be of the form $\mathrm{diag}(-x, x)$. Combining these facts, the vacuum stress tensor, which is invariant under 3 boosts and 3 rotations, must be of the form $\mathrm{diag}(-x, x, x, x)$. Recalling that an observer with zero 3-velocity observes the stress tensor as energy density: $\rho_\Lambda = T_{\mu\nu} u^\mu u^\nu$. Thus the vacuum stress tensor has to be

$$T^\mu{}_\nu = \mathrm{diag}(-\rho_\Lambda, p_\Lambda, p_\Lambda, p_\Lambda), \qquad p_\Lambda = -\rho_\Lambda. \tag{2.6}$$

In other words, the vacuum stress tensor only has one free parameter to determine: the energy density ρ_Λ.

Up to now, we have only considered Lorentz symmetry, which can be defined locally at a point in spacetime. Thus ρ_Λ could still be a function of spacetime point x^μ. Spacetime translation symmetry will further reduce the available possibilities. For Minkowski space (i.e. $g_{\mu\nu} = \mathrm{diag}(-1, 1, 1, 1)$), there are 4 translation symmetries thus ρ_Λ must be a constant independent of spacetime. For flat FRW spacetime (i.e. the metric (1.13)), time translation symmetry is broken thus the most general choice[4] for vacuum stress tensor is $\rho_\Lambda = \rho_\Lambda(t)$.

---

[4] Here we restrict our attention to the case that the FRW spacetime is a good description of our universe. For some dynamical dark energy models, for example scalar field models of dark energy, spatial structure in the universe could also affect dark energy density, thus in those cases ρ_Λ may be a general function $\rho_\Lambda(x)$.

So far we have considered the form of the vacuum energy for an inertial observer. For an arbitrary (i.e. can accelerate) observer u^μ, the energy-momentum tensor changes its form under general coordinate transformation, and $T_{\mu\nu} = (\rho_\Lambda + p_\Lambda)u^\mu u^\nu + p_\Lambda g_{\mu\nu} = -\rho_\Lambda g_{\mu\nu}$.

Before ending up this section, let us mention one more important thing: As a quantum analog of "nothing", the vacuum state can be on the one hand considered as an energy component of the system; and also on the other hand considered as a part of the definition of a theory. In short, it can be considered either as a state or a law. Applying to the Einstein equations, we can write the vacuum stress tensor as $T_{\mu\nu} = -\rho_\Lambda g_{\mu\nu}$, and move this term to the left hand side of the Einstein equations. As a result, we have

$$R_{\mu\nu} - \frac{1}{2}g_{\mu\nu}\mathcal{R} + 8\pi G \rho_\Lambda g_{\mu\nu}$$

$$= 8\pi G \times \text{(stress tensor from other things)}. \qquad (2.7)$$

Define $\Lambda \equiv 8\pi G \rho_\Lambda$, the vacuum energy density (at best when it is a constant) could be thought of a parameter in the theory. This is actually the form when the vacuum energy first appears in theoretical physics, known as the cosmological constant, introduced by Einstein.

2.5 Particles

Here we describe particles in the language of fluids.

For non-relativistic particles, the kinetic energy of a particle is much smaller than the static energy, thus the energy interchange under collision (thus pressure) is much smaller than the static energy (thus energy density).[5] In other words, we have $p \ll \rho$. To the first order approximation, we can set $p = 0$ and $T_{\mu\nu} = \rho u_\mu u_\nu$. The equation of state is $w = 0$. When the physics we are interested in nontrivially span over a large variety of scales, we may need consider the small pressure as well. In those cases the sound speed is given by[6] $c_s = \sqrt{w} \equiv \sqrt{p/\rho}$.

---

[5] As a familiar example, for the air around us on the earth, we have velocity $v \sim 10^{-7}$, and $p/\rho \sim 10^{-12}$.

[6] To derive c_s here, one has to insert a small perturbation to the perfect fluid stress tensor, and solve the conservation equation with an equation of state $p = p(\rho)$. Thus the sound speed given by $\sqrt{w}$ only applies with the above limitations. Especially, for the cosmological constant, although $w = -1$, there is no perturbation thus no definition of sound speed. For a scalar field, on sub-Hubble scales p is independent of ρ, thus $c_s \neq \sqrt{w}$, and on super-Hubble scales it behaves like a cosmological constant.

For extreme relativistic particles, whose energy is almost (or all) made up of kinetic energy, the dispersion relation of such a particle is $\omega = k$. The partition function can be obtained by inserting $\omega = k$, and the equation of state can be calculated from the partition function. We shall not do it explicitly here. The result is that the ratio between pressure and energy density is fixed: $p = \rho/3$. The ratio can also be heuristically understood as an average of motion of massless particles in three spatial dimensions. The speed of sound in this case is $c_s = 1/\sqrt{3}$, although an individual particle moves at the speed of light $c = 1$, the majority does not move in the same direction.

2.6   Homogeneous Field Configurations

Energy can also be carried by homogeneous field configurations. Here we restrict our discussion to the simplest possible field: a scalar field with standard kinetic term. The action of such a scalar field is

$$S = \int d^4 x \sqrt{-g} \left(-\frac{1}{2} g^{\mu\nu} \partial_\mu \phi \partial_\nu \phi - V(\varphi) \right). \tag{2.8}$$

Recall that our metric convention is $(-, +, +, +)$. Thus the time derivative term in the action has positive coefficient.

The stress tensor can be worked out from Eq. (1.23):

$$T_{\mu\nu} = \partial_\mu \phi \partial_\nu \phi - g_{\mu\nu} \left(\frac{1}{2} g^{\rho\lambda} \partial_\rho \phi \partial_\lambda \phi + V(\phi) \right). \tag{2.9}$$

To compare this result with the fluid stress tensor (2.2), we have

$$\rho = \left(-\frac{1}{2} g^{\rho\lambda} \partial_\rho \phi \partial_\lambda \phi + V(\phi) \right), \qquad p = \left(-\frac{1}{2} g^{\rho\lambda} \partial_\rho \phi \partial_\lambda \phi - V(\phi) \right), \tag{2.10}$$

$$u_\mu = -\frac{\partial_\mu \phi}{\sqrt{-g^{\rho\lambda} \partial_\rho \phi \partial_\lambda \phi}}. \tag{2.11}$$

Note that the pressure is identical to the Lagrangian density.

Up to here, the result is completely general. For the purpose of cosmology, we are going to consider a spatially homogeneous field configuration. Thus we let $\varphi = \varphi(t)$. Also we take the spatial flat FRW metric (1.13). In this case we have

$$\rho = \frac{1}{2} \dot{\phi}^2 + V, \qquad p = \frac{1}{2} \dot{\phi}^2 - V,$$
$$u_\mu = (-1, 0, 0, 0), \qquad T_{\mu\nu} = \mathrm{diag}(\rho, p, p, p). \tag{2.12}$$

Fluctuations around this homogeneous background $\phi(x) = \phi_0(t) + \delta\phi(x)$, $g_{\mu\nu}(x) = g_{\mu\nu}^{(0)}(t) + \delta g_{\mu\nu}(x)$ is also important in cosmology. We are not going into those details here.

3
Cosmology

3.1 The Cosmological Principle

Our universe is extremely large and complicated. It is hopeless to follow each object in the universe and construct a description of the whole universe on top of that. To setup a theory for cosmology, some levels of approximations are essentially required. The cosmological principle is what we need for this purpose.

The cosmological principle asserts that, on large scales,[1] our observable universe can be well approximated as homogeneous and isotropic. The principle is first imposed by Einstein, and as a theoretical approximation because experimental data is not well enough to test the cosmological principle at that time. Surprisingly, decades later, data turn out to support the cosmological principle as precisely as one part in $10^4 \sim 10^5$. This fact recalls us Einstein's saying: "The most incomprehensible thing about the world is that it is comprehensible."

As the cosmological principle is extremely important in modern cosmology, for a clear understanding, several comments are in order:

• Homogeneity means that there is no particular points in the universe, and isotropy means that at one point, there is no particular direction. Isotropy at every spatial point implies homogeneity. However, homogeneity for each spatial point does not imply isotropy because one can imagine a homogeneous universe with a global gradient.

• As in numerous other areas of physics, once a background model is setup, one can do perturbation theory around it. In cosmology it is exactly the case. The deviation from the cosmological principle can be treated as

---

[1] To be explicit, the size of our observable universe is about 3000 Mpc. On scales greater than about 100 Mpc, the universe is approximately homogeneous and isotropic.

perturbations and calculated order by order. Compared with the homogeneous and isotropic background, a much greater amount of information about cosmology is stored in the perturbations.

• Although the cosmological principle is nowadays well tested in our observable universe, it is dangerous to assume a priori that the same holds on much greater scale than our Hubble radius. For example, eternal inflation, as a popular theory for the very early universe, produces an extremely inhomogeneous state on super-Hubble length scales.

3.2 Newtonian Cosmology

As a quick start, cosmology can be intuitively (but in some sense inaccurately) understood in terms of Newton mechanics. We start from a matter dominated universe. Consider a ball with radius $R(t)$ in the universe. The gravitational force (per unit mass) acting on the boundary of the ball is

$$\ddot{R} = -\frac{GM}{R^2} = -\frac{4\pi G}{3}\rho R, \tag{3.1}$$

where M is the mass inside the ball, related to energy density by $M = 4\pi R^3 \rho/3$. To be a little bit more formal, the above equation can also be obtained from the Poisson equation (1.2).

Also, from energy conservation, we have

$$\frac{1}{R^3}\frac{d(\rho R^3)}{dt} = \dot{\rho} + 3\frac{\dot{R}}{R}\rho = 0. \tag{3.2}$$

From Eqs. (3.1) and (3.2), we also have

$$\left(\frac{\dot{R}}{R}\right)^2 = \frac{8\pi G}{3}\rho. \tag{3.3}$$

Thus once matter is initially "thrown" out by some big bang, the particles will be flying outwards with a negative acceleration. Equation (3.3) is essentially the well known Friedmann equation, with replacement of ball radius $R(t)$ by scale factor $a(t)$. The replacement indicates that the conception that particles are flying outwards is to be replaced by the expansion of space in general relativity.

Again, fluctuations on top of the homogeneous background can also be considered in Newtonian cosmology, but beyond the scope of our purpose here.

Before concluding this section, we would like to mention there are cautions that the Newtonian cosmology must be carefully used or interpreted. For the following reasons, Newtonian cosmology is not trustable and a relativistic description is needed:

The universe is large and mass sums up. In the case of a Schwarzschild black hole, it is well known that Newtonian gravity breaks down near the black hole horizon, where the radius is of order GM (where M is the mass of the black hole). The observable universe, though different from a black hole, also has a radius of order GM, where M is now the total mass inside the observable universe. Thus there is no reason to fully trust Newtonian gravity on such large scales.

Matter used to be relativistic (radiation) in the early universe. Newtonian cosmology cannot deal with such relativistic matter by definition. An explicit way to see this is that in the Poisson equation, only energy density is taken into consideration. However, pressure also plays an important role in the stress tensor. To treat pressure correctly, relativity is required.

Newtonian cosmology also faces a number of conceptual problems, which are solved by general relativity. For example, to solve the Poisson equation in infinite space, boundary conditions at spatial infinity need to be imposed. However, matter does not vanish at infinity thus the boundary condition may not be trivial. To put this problem in physics context: to have the simplest setup, we have implicitly assumed that the universe outside the ball $R(t)$ is spherical and take the radius of the sphere to infinity. In this way the matter outside the ball $R(t)$ does not have gravitational effect for the inside matter. However, if we have chosen an alternative shape of matter outside the ball before taking the limit, we get a different result. This problem is solved by general relativity because a relativistic theory of gravity is causal, thus does not rely on a boundary condition at spatial infinity.

3.3 FRW Cosmology

In this section, we setup some basics of cosmology in the framework of general relativity. The starting point is restrictions on the metric $g_{\mu\nu}$. As discussed in Section 3.1, we approximate our universe to be homogeneous and isotropic. The most general geometry that fits those symmetries can be described by the metric

$$ds^2 = -dt^2 + a(t)^2 \left[\frac{dr^2}{1 - kr^2} + r^2 \left(d\theta^2 + \sin^2\theta d\varphi^2 \right) \right]. \qquad (3.4)$$

This metric is known as the Friedmann–Robertson–Walker (FRW) metric (also called the Friedmann–Lemaître–Robertson–Walker metric).

Here k is a parameter that not determined by the symmetry. Different choice of k results in different geometries. The simplest choice is $k = 0$. In this case, the metric reduces to the metric (1.13):

$$ds^2 = -dt^2 + a(t)^2 dx^i dx^i \,. \tag{3.5}$$

Here the spatial sections (equal time hyper-surfaces) are flat.

Note that under the redefinition $a(t) \to \lambda a(t)$, $x^i \to x^i/\lambda$, the metric keeps its form invariant. Thus neither $a(t)$ or x^i has any physical meaning by themselves. Another way to see this is that there is no scale on a flat spatial section, thus the "scale" factor on a single section is not meaningful. On the other hand, the combination $a(t)x^i$ has the meaning of physical distance measured by a comoving observer. Also, the ratio of scale factors at different times: $a(t_1)/a(t_2)$ is meaningful, which plays a key role in cosmology.

When $k \neq 0$, by rescaling $a(t) \to \lambda a(t)$, $r^i \to r^i/\lambda$, $k \to \sqrt{\lambda}k$, the metric is kept invariant. Using this rescaling, we can always rescale k to be either $+1$ or -1. These two choices corresponds to spherical (positively curved) and hyperbolic (negatively curved) spatial sections respectively. In these cases, we have already used the rescaling to fix k. Thus the scale factor becomes physically meaningful: $a(t)$ is the curvature radius of the universe.

Until now, the $k = 0$ case agrees with experiments very well. Thus in later chapters we will focus on this case unless mentioned otherwise. Even if we detect any deviation from $k = 0$ in the future, the difference has to be small (within several percent), thus $k = 0$ is anyway a good approximation of our universe.

Insert the flat FRW metric (3.5) to the Einstein equation (1.19), we get (after linear combination of the equations)

$$3M_{\rm P}^2 H^2 = \rho \,, \tag{3.6}$$

$$-2M_{\rm P}^2 \dot{H} = \rho + p \,. \tag{3.7}$$

These equations (or sometimes specialized to the first equation) are called the Friedmann equations. Note that in deriving these equations, for $T_{\mu\nu}$ we have chosen the frame $u^\mu = (1, 0, 0, 0)$ from symmetry consideration.

As one can read from the first Friedmann equation, energy density makes the universe change its volume. The change may be either expansion or contraction. Our present universe is expanding. Thus we will choose the expansion direction unless otherwise mentioned.

Pressure affects acceleration. To see it more explicitly, it is useful to combine Eqs. (3.6) and (3.7) into

$$\frac{\ddot{a}}{a} = -\frac{\rho + 3p}{6M_{\rm P}^2} \,. \tag{3.8}$$

Thus when $p > -\rho/3$, the universe decelerates. It is convenient to define a "deceleration parameter" $q \equiv -\ddot{a}a/\dot{a}^2$. When $q > 0$, the universe decelerates. On the other hand, when $p < -\rho/3$, the universe accelerates.

The various forms of the stress tensor discussed in Chapter 2 can be discussed here one by one. We summarize the result here. For an energy component satisfying an equation of state $p = w\rho$ with a constant w, the scale factor a can be solved as

$$\frac{a(t_2)}{a(t_1)} = \left(\frac{t_2}{t_1}\right)^{\frac{2}{3(1+w)}} \qquad (w \neq -1) \,. \tag{3.9}$$

Especially, for radiation domination, $a \propto \sqrt{t}$. For matter domination, $a \propto t^{2/3}$: less pressure makes the universe lose energy more slowly and thus expand more quickly. When $w = -1$, the exponent on the right hand side blows up, which indicates that the polynomial turns to an exponential. Indeed, by solving the Friedmann equations for $w = -1$, one gets

$$\frac{a(t_2)}{a(t_1)} = e^{H(t_2 - t_1)} \,. \tag{3.10}$$

In astrophysics and cosmology, it is convenient to use the redshift parameter z to measure distance. As photons redshift as $1/a(t)$, the redshift parameter is related to scale factor as

$$a = 1/(1+z) \,, \tag{3.11}$$

where we have set $a(t_0) = 1$.

In addition, it is convenient to introduce the fractional density of matter, radiation, and dark energy

$$\Omega_{\rm m} \equiv \frac{\rho_{\rm m}}{\rho_{\rm c}} \,, \qquad \Omega_{\rm r} \equiv \frac{\rho_{\rm r}}{\rho_{\rm c}} \,, \qquad \Omega_{\rm de} \equiv \frac{\rho_{\rm de}}{\rho_{\rm c}} \,, \tag{3.12}$$

where $\rho_{\rm c} \equiv 3M_{\rm p}^2 H^2$ is the critical density. Notice that these quantities are functions of redshift z. In the following context, we will use $\Omega_{\rm m0}$, $\Omega_{\rm r0}$, $\Omega_{\rm de0}$, and $\rho_{\rm c0}$ to denote their present-day values.

Part II

Theoretical Aspects

4
Introduction to Dark Energy

4.1   The Cosmological Constant Reloaded

Since its discovery in 1998 [1, 2], dark energy has become one of the central problems in theoretical physics and cosmology [3–14] Thousands of papers have been written on this subject, while it is still as daunting as ever to understand the nature of dark energy.

We start with a brief history of the dark energy problem in chronological order [3, 15].

1917: Einstein added a cosmological constant term in his field equations, for the following reasons: firstly, for isolated mass not to impose a structure on space at infinity in a closed universe; secondly to obtain a static universe [16].

1920s: Pauli realized that for a radiation field the vacuum energy is too large to gravitate. As we shall review, the vacuum energy density of a radiation field is proportional to the cutoff to the fourth power. Pauli (unpublished) showed that using the classical electron radius as an ultraviolet cutoff, the curvature of the universe will be so large that the universe "could not even reach to the moon".

1931: Einstein removed the cosmological constant [17] because of the discovery of the cosmic expansion. Although in 1923 on a postcard to Weyl he already wrote: "If there is no quasi-static world, then away with the cosmological term!" It seems that he did not believe in "no quasi-static world" until 1931 [17].

1960s: To explain why there are so many quasars centering around redshift around $z = 1.95$, some people suggested to use the Lemaître model (with $\Lambda > 0$, $k = 1$), and around $a = 1/(1 + z) = 1/2.95$. In their model, the universe is approaching the Einstein's static universe near $z = 1.95$, with a positive cosmological constant and a positive curvature [18], and after a while starts to expand again because Einstein's static universe is an unstable solution.

1967: Zel'dovich reintroduced the cosmological constant problem by taking the vacuum fluctuations into account. This introduced the old cosmological constant problem, Zel'dovich used the word "fine-tuning" [19, 20].

1987: Weinberg "predicted" a non-vanishing and small cosmological constant [21], and two years later published the by now famous review article [3].

1998: Based on the analysis of 16 distant and 34 nearby supernovae, Riess *et al.* first discovered the acceleration of expanding universe [1]. Soon after, utilizing 18 nearby supernovae from the Calan–Tololo sample and 42 high-redshift supernovae, Perlmutter *et al.* confirmed the discovery of cosmic acceleration [2].

2000s: String theorists reintroduced the anthropic principle when discovered the string landscape [22, 23].

2006: Dark Energy Task Force proposed a quantitative "figure of merit" to characterize the performance of dark energy survey projects [24].

2011: Because of the discovery of cosmic acceleration, Saul Perlmutter, Brian Schmidt, and Adam Riess won the Nobel prize in physics 2011.

4.2   The Theoretical Challenge

Einstein was the first to introduce the famous cosmological constant term in his equations. In the Einstein equation

$$R_{\mu\nu} - \frac{1}{2}g_{\mu\nu}\mathcal{R} = 8\pi G T_{\mu\nu} \,, \tag{4.1}$$

one adds a term to the stress tensor on the R.H.S. of the above equation

$$T_{\mu\nu} \rightarrow T_{\mu\nu} - \frac{1}{8\pi G}\Lambda g_{\mu\nu} \,, \tag{4.2}$$

where Λ is a constant, the cosmological constant. The Einstein equation can be rewritten as

$$R_{\mu\nu} - \frac{1}{2}g_{\mu\nu}\mathcal{R} + \Lambda g_{\mu\nu} = 8\pi G T_{\mu\nu} \,. \tag{4.3}$$

Before trying to understand the nature of Λ, we note that it has a couple of ready interpretations. First, if we take the additional term in Eq. (4.2) as coming from some ideal fluid, whose stress tensor is given by

$$T_{\mu\nu} = (\rho_{\text{de}} + p_{\text{de}})u_\mu u_\nu + p_{\text{de}}g_{\mu\nu}, \tag{4.4}$$

where ρ_{de} and p_{de} are the energy density and the pressure of dark energy, respectively.

Then the cosmological constant can be interpreted as a fluid with

$$p_{\text{de}} = -\rho_{\text{de}}, \quad \rho_{\text{de}} = \frac{1}{8\pi G}\Lambda, \tag{4.5}$$

and the equation of state is $w_{\text{de}} = -1$. This is certainly an unusual fluid, since if $\Lambda > 0$, the pressure is negative, and the strong energy condition is violated since $(T_{\mu\nu} - \frac{1}{2}Tg_{\mu\nu})u^\mu u^\nu = -\Lambda < 0$. Of course the strong energy condition is not anything sacred. The null energy condition is marginally satisfied.

Second, since the cosmological constant term is proportional to $g_{\mu\nu}$ thus is Lorentz invariant, it can be interpreted as the vacuum energy. And indeed Lorentz invariance forces upon us the condition $p_\Lambda = -\rho_\Lambda$. In principle, before we understand the origin of the vacuum energy, the energy density can be positive, negative and zero.

In the Friedmann–Robertson–Walker cosmology, the two Friedmann equations read

$$3M_{\text{P}}^2 H^2 = \rho_{\text{m}} + \rho_{\text{de}} - \frac{3M_{\text{P}}^2 k}{a^2},$$
$$6M_{\text{P}}^2 \frac{\ddot{a}}{a} = 2\rho_{\text{de}} - \rho_{\text{m}}, \tag{4.6}$$

where $H = \dot{a}/a$ is the Hubble "constant", ρ_{de} is the vacuum energy density $\rho_{\text{de}} = M_{\text{P}}^2\Lambda$, and ρ_{m} is the matter energy density (with $p_{\text{m}} = 0$). Einstein introduced a positive cosmological constant motivated by a static universe. To have a static universe $H = 0$ and without the cosmological constant, we deduce from the first equation above $\rho_{\text{m}} = 3M_{\text{P}}^2/a^2$ for a positive spatial curvature $k = 1$. But this does not work, since the second Friedmann equation we have $\ddot{a}/a < 0$ for ρ_{m}. Namely the universe will collapse due the attractive force of matter. Thus, to have both $H = 0$ and $\ddot{a}/a = 0$ we need to have $2\rho_{\text{de}} = \rho_{\text{m}}$ and $\rho_{\text{de}} = M_{\text{P}}^2/a^2$. This is why the cosmological constant has to be positive and fine-tuned to balance matter. The spatial curvature also has to be positive.

However, due to the cosmic expansion, Einstein himself later abandoned the cosmological constant. Nevertheless, as the cosmological constant is the simplest extension of the original Einstein equations, the theoretical possibility is left open.

Up to now, we have only discussed the classical story. Without quantum mechanics, a very small cosmological constant poses no problem, one simply regards Λ as a parameter in theory. However, in quantum mechanics, we know that vacuum fluctuations make contribution to the energy of a vacuum, thus the vacuum energy density receives two contributions, one may be called the bare vacuum energy, the classical one without quantum contribution, the other comes from the zero-point fluctuations of all quantum fields (in quantum field theory).

For a quantum field with a given mode of frequency ω, the zero-point energy is $\pm\frac{1}{2}\omega$, with the plus sign for a bosonic field and the minus sign for a fermionic field. The total zero-point energy is then $\frac{1}{2}\sum_i(\pm)\omega_i$. In the continuum limit, we have for a free field

$$
\frac{1}{2}\sum_i \omega_i = \frac{1}{2}\int \frac{d^3x d^3k}{(2\pi)^3}(k^2+m^2)^{\frac{1}{2}}
$$

$$
= V\int \frac{k^2 dk}{4\pi^2}(k^2+m^2)^{\frac{1}{2}}, \tag{4.7}
$$

this integral is divergent, thus we need to introduce a physical cutoff. A cutoff is reasonable, for example, if the zero-point energy density is infinite, then our universe is infinitely curved and the space size is infinitely small. Thus by examining the Friedmann equations, we know that this cutoff must be the Planck energy at largest.

Let the cutoff be $\lambda \gg m$, the energy density of a bosonic field is then

$$
\rho_{\text{boson}} = \int_0^\lambda \frac{k^2 dk}{4\pi^2}(k^2+m^2)^{\frac{1}{2}} \approx \int_0^\lambda \frac{k^3 dk}{4\pi^2} = \frac{\lambda^4}{16\pi^2}. \tag{4.8}
$$

Take, for example, $\lambda = M_{\text{P}}$, then

$$
\rho_{\text{boson}} = \frac{1}{2^{10}\pi^4 G^2} \approx 2\times 10^{71}\ \text{GeV}^4.
$$

But in reality, $\rho_{\text{de}} < \rho_{\text{c}} \approx (3\times 10^{-12}\ \text{GeV})^4 \approx 10^{-46}\ \text{GeV}^4$. If there is no supersymmetry, $|\rho_\Lambda|$ should be greater than ρ_{boson} from a single bosonic field, namely $\rho_{\text{boson}} < \rho_{\text{c}}$, but what we have is

$$\frac{\rho_{\text{boson}}}{\rho_{\text{c}}} \approx 10^{117}. \tag{4.9}$$

This is the famous cosmological constant problem. Because if the cosmological constant is much larger than ρ_{c}, the universe will never look like what we observe today.

With exact supersymmetry, the bosonic contribution to cosmological constant ρ_{boson} is canceled by its fermionic counterpart. However, we know that our world looks not supersymmetric. Supersymmetry, if exists, has to be broken above or around 100 GeV scale. Even if we take the $\Lambda_{\text{SM}} = 100$ GeV cutoff, below which physics is believed to be well-described by the particle physics standard model, still

$$\rho_{\text{boson}} \approx \frac{\Lambda_{\text{SM}}^4}{16\pi^2} = 10^6 \text{ GeV}^4, \qquad \frac{\rho_{\text{boson}}}{\rho_{\text{c}}} \approx 10^{52}. \tag{4.10}$$

One could use fine-tuning to solve the problem in some sense, by introducing a bare cosmological constant and letting it cancel with the quantum contribution using renormalization. However, one has to make two independent numbers cancel by the accuracy of one part in 10^{117} or at least 10^{52}. This is extremely unlikely to happen. Thus it remains a problem why the cosmological constant is not large. This is known as the old cosmological constant problem [19].

The cosmological constant problem remained the above statement until crucial experiments came in. Riess *et al.* and Perlmutter *et al.* in 1998 discovered the accelerating expansion of our universe [1, 2]. The simplest explanation of this phenomenon is the return of a positive cosmological constant. Let $\rho_{\text{c}} = 3M_{\text{p}}^2 H^2$ be the critical energy density, data available today tell us that $\rho_{\text{de}} = 0.73\rho_{\text{c}}$ and $\rho_{\text{m}} = 0.27\rho_{\text{c}}$. Thus not only there is a small positive cosmological constant, but also its energy density is the same order of matter energy density. These discoveries lead to a new version of the cosmological constant problem, or the problem of dark energy.

The modern version of the cosmological constant problem divides itself into two parts:

(a) Why $\rho_{\text{de}} \approx 0$, namely why it is so small? Sometimes, this problem is also split into two: (a-1) Why the cosmological constant is so small, not at Planck scale or at least GeV scale? (a-2) Why the cosmological constant is non-zero? Because a small number may be zero for some reasons, a small but non-zero number is more curious.

(b) Why $\rho_{de} \sim \rho_m$ now? This is called the coincidence problem. Because if ρ_{de} were indeed a cosmological constant, $\rho_{de} \sim \rho_m$ happens only during a tiny fraction of energy scales, comparing with the whole history and future of our universe.[1]

Our writing of the theoretical part of this book is motivated by Weinberg's review article [3], we will first review the theoretical efforts before the discovery of acceleration [1, 2], following Weinberg's classification:

1. Supersymmetry and superstring.

2. The anthropic principle.

3. The tuning mechanisms.

4. Modifying general relativity.

5. Ideas of quantum gravity.

Twenty years have passed since Weinberg's classification, all the new ideas more or less still belong to the above five categories. For example, category 1 and category 2 combine to become the string landscape + the anthropic principle, and there are new ideas in the category of symmetries. The tuning mechanisms now include ideas associated with brane-world models. The category of modifying gravity now includes $f(\mathcal{R})$ models, $f(\mathcal{T})$ models, MOND and TeVeS models and the Dvali–Gabadadze–Porrati (DGP) model. There are also some progress in the category of quantum gravity.

Further, we can add three more categories: the holographic principle, back-reaction of gravity, phenomenological models. Thus, we will write about:

1. Symmetry.

2. The anthropic principle.

3. The tuning mechanisms.

4. Modifying gravity.

5. Quantum gravity.

6. The holographic principle.

7. Back-reaction of gravity.

8. Phenomenological models.

---

[1] On the other hand, even dark energy is dynamical and will remain today's value in the future, the coincidence problem is still not solved. Because still we need to explain why we are living in the epoch that dark energy just starts to dominate, instead of living in a universe that $\rho_\Lambda \sim \rho_m$ for a very long time.

References

[1] A. G. Riess *et al.*, AJ. **116** (1998) 1009.

[2] S. Perlmutter *et al.*, ApJ. **517** (1999) 565.

[3] S. Weinberg, Rev. Mod. Phys. **61** (1989) 1.

[4] S. M. Carroll, W. H. Press, and E. L. Turner, Ann. Rev. Astron. Astrophys. **30** (1992) 499.

[5] T. Padmanabhan, Phys. Rept. **380** (2003) 235.

[6] P. J. E. Peebles and B. Ratra, Rev. Mod. Phys. **75** (2003) 559.

[7] E. J. Copeland, M. Sami, and S. Tsujikawa, Int. J. Mod. Phys. D **15** (2006) 1753.

[8] J. P. Uzan, Gen. Rel. Grav. **39** (2007) 307.

[9] E. V. Linder, Rept. Prog. Phys. **71** (2008) 056901.

[10] J. Frieman, M. Turner, and D. Huterer, Ann. Rev. Astron. Astrophys. **46** (2008) 385.

[11] R. Durrer and R. Maartens, arXiv:0811.4132.

[12] S. Tsujikawa, arXiv:1004.1493.

[13] Y. Wang, *Dark Energy* (Wiley-VCH, 2010).

[14] V. Sahni, A. Starobinsky, Int. J. Mod. Phys. D **9** (2000) 373; S. M. Carroll, Living Rev. Rel. **4** (2001) 1; V. Sahni, Lect. Notes. Phys. **653** (2004) 141; J. A. S. Lima, BJP, **34** (2004) 1; D. H. Weinberg, New. Astron. Rev. **49** (2005) 337; N. Straumann, Mod. Phys. Lett. A **21** (2006) 1083; M. S. Turner, D. Huterer, J. Phys. Soc. Jap. **76** (2007) 111015; J. P. Uzan, arXiv:0912.5452; A. Silvestri, M. Trodden, Rept. Prog. Phys. **72** (2009) 096901; R. P. Caldwell, M. Kamionkowski, Ann. Rev. Nucl. Part. Sci. **59** (2009) 397; P. Peter and J. P. Uzan, *Primordial Cosmology* (Oxford University Press, Oxford, 2009); D. Sapone, Int. J. Mod. Phys. A **25** (2010) 5253.

[15] N. Straumann, arXiv:gr-qc/0208027.

[16] A. Einstein, Sitz. Preuss. Akad. Wiss. Phys-Math **142** (1917) 87.

[17] A. Einstein, Sitz. Preuss. Akad. Wiss. Phys-Math **235** (1931) 37.

[18] V. Petrosian, E. Salpeter, and P. Szekeres, ApJ. **147** (1967) 1222.

[19] Y. B. Zel'dovich, JETP Lett. **6** (1967) 316.

[20] V. Sahni, A. Krasinski, and Y. B. Zel'dovich, Sov. Phys. Usp. **11** (1968) 381.

[21] S. Weinberg, Phys. Rev. Lett. **59** (1987) 2607.

[22] R. Bousso and J. Polchinski, JHEP **0006** (2000) 006.

[23] L. Susskind, arXiv:hep-th/0302219.

[24] A. Albrecht *et al.*, arXiv:astro-ph/0609591.

5
Weinberg's Classification

As mentioned, before the accelerating expansion of the universe was observed, theorists have already been worrying about the cosmological constant. In this chapter, we will briefly review Weinberg's classification of ideas about the cosmological constant problem [1], before the discovery of dark energy.

5.1 Supersymmetry

In any supersymmetric theory in 4 dimensions, the supersymmetry algebra contains at least $\mathcal{N} = 1$ generators Q_α and their conjugate $Q_\alpha^\dagger$ [2] such that

$$\{Q_\alpha, Q_\beta^\dagger\} = (\sigma_\mu)_{\alpha\beta} P^\mu \,, \tag{5.1}$$

where σ_i are the Pauli matrices and $\sigma_0 = 1$. If supersymmetry is unbroken, then

$$Q_\alpha|\Omega\rangle = Q_\alpha^\dagger|\Omega\rangle = 0 \,, \tag{5.2}$$

the supersymmetry algebra leads to

$$H|\Omega\rangle = P^0|\Omega\rangle = 0 \,, \tag{5.3}$$

namely, the unbroken vacuum has exactly zero energy [3].

In a supersymmetric quantum field theory, there are a number of chiral superfields for which the potential V is determined by the superpotential W. W is a function of complex scalar fields ϕ^i in the chiral multiplets,

$$V = \sum_i |\partial_i W|^2 \,, \tag{5.4}$$

where $\partial_i W = \partial W/\partial \phi^i$. For a vacuum with unbroken supersymmetry $V = 0$ thus $\partial_i W = 0$.

Of course any supersymmetry must be broken in our world, thus in general $\sum_\alpha \{Q_\alpha, Q_\alpha^\dagger\} = 2H > 0$. If there is translation symmetry in spacetime, we must have $H \sim V\rho_{\mathrm{de}}$ thus $\rho_{\mathrm{de}} > 0$. So a positive cosmological constant is a consequence of broken supersymmetry in quantum field theory, as long as this is the whole contribution to the effective cosmological constant. This conclusion agrees with observations since 1998, but what is the exact value of ρ_{de}? For a typical quantum field theory, one expects $\rho_{\mathrm{de}} \sim M_{\mathrm{SUSY}}^4 \gg \rho_{\mathrm{c}}$. Thus we would say that supersymmetry does not solve the cosmological constant problem.

It was commonly thought that it is better to explain $\rho_{\mathrm{de}} = 0$ first, then take the next step to explain why $\rho_{\mathrm{de}} \sim \rho_{\mathrm{c}}$.

The potential in a supergravity theory is determined by both the super-potential and the Kähler potential $K(\phi^i, \bar{\phi}^i)$ [4],

$$V = e^K \left[G^{i\bar{j}} D_i W \overline{D_j \phi} - 3|W|^2 \right], \qquad (5.5)$$

where we already set $8\pi G = 1$, and

$$D_i W = \partial_i W + \partial_i K W, \qquad G_{i\bar{j}} = \partial_{\phi^i} \partial_{\bar{\phi}^j} W, \qquad G^{i\bar{j}} G_{\bar{j}k} = \delta_k^i. \qquad (5.6)$$

In a class of the so-called no-scale supergravity models [5], one can fine-tune parameters to break supersymmetry, meanwhile keep the vacuum energy vanishing. There are three classes of complex scalar fields, C^a, S^n and T.

Let

$$K = -3\ln(T + \bar{T} - h(C, \bar{C})) + \tilde{K}(S, \bar{S}), \qquad W = W_1(C) + W_2(S), \qquad (5.7)$$

then

$$V = e^{\tilde{K}} \left[(T + \bar{T} - h)^{-3} N^{a\bar{b}} \partial_a W \overline{\partial_b W} + G^{m\bar{n}} D_m W \overline{D_n W} \right], \qquad (5.8)$$

where $N = \partial_a \partial_{\bar{b}} h$. Since both N and G are positive definite, so $V \geq 0$. To have $V = 0$ there ought to be

$$\partial_a W = D_m W = 0. \qquad (5.9)$$

The point where $V = 0$ is a minimum, so there is no instability problem. For

$$D_a W = \partial_a W + \partial_a K W = \partial_a K W \sim \partial_a h W \,, \qquad (5.10)$$

if $W \neq 0$, then $D_a W \neq 0$ and by definition supersymmetry is broken.

In such a model, T is not fixed, this is why a model like this is called no-scale supergravity model. The problems for this model include:

- The coefficient of the first term must be -3, this is fine-tuning.
- The form of W is fine-tuned.
- Quantum corrections usually spoil those fine-tuned coefficients.

The no-scale supergravity models continue to attract attention today.

5.2   Anthropic Principle

The terminology of the anthropic principle is due to Brandon Carter, who articulated the anthropic principle in reaction to the Copernican Principle, which states that humans do not occupy a privileged position in the Universe. Carter said: "Although our situation is not necessarily central, it is inevitably privileged to some extent." [6]

As Weinberg formulated in his review [1], there are three different kinds of anthropic principle: very strong version, very weak version and weak version.

The very strong version states that everything in our universe has something to do with humankind, this is of course absurd.

The very weak version takes the very existence of our humankind as a piece of experimental data. For instance, in order not to kill a person with the products of radio-decay, the lifetime of a proton must be at least 10^{16} years. The very weak version of anthropic principle is as correct as any other well made experiments, but not useful. Because one can always design experiments with better (and often much better) precision than "not to kill a person". For example, proton decay experiments have put a lower bound on proton lifetime of order 10^{29} years (or 10^{33} years with assumptions on decay product), much longer than the very weak anthropic bound. At the very least the person will feel sick (as kind of experiment in a generalized sense) before being killed.

Now the weak version. This is the version that prevails in certain circle of people. In this version, it is postulated that there are many regions in the universe. In these regions physical laws are in different forms. It just happens that in the region we are dwelling, all physical laws, physical constants and cosmological parameters are such that clusters of galaxies, galaxies and our solar system can form, and humankind can appear. It appears that all the conditions are fine-tuned but they are not, because all different sorts of regions exist in the universe, such a universe is called multiverse. We found ourselves in our part of the multiverse simply because this is all we can observe.

In such a weak version, we are not supposed to expect every physical law and every physical constant be designed for the existence of humankind. For instance, we do not know whether the fact that the lifetime of proton is more than 10^{29} years has anything to do with us, or whether it is connected with other conditions which are necessary for the existence of us.

Dicke [7, 8] may be one of the first persons making use of this weak version of the anthropic principle, when he was considering Dirac's large number problem. Let t_{age} be the age of the universe. The death date of the sun is greater than t_{age}, and t_{age} is greater than the formation time of the second and the third generations of stars. Thus t_{age} is about 10^{10} years. This is a nice explanation of a large number. However, other time scales such as $1/(m_\pi^3 l_P^2) \sim 10^{10}$ years are completely incomprehensible in this way.

Weinberg's anthropic consideration of the cosmological constant [9] is considered to be a genuine prediction of the anthropic principle by some people. The simplified version of his argument is the following. Let z be the redshift when galaxies form, the matter density is $\rho_m(z) = (1 + z)^3 \rho_{m0} \sim 100\rho_{m0}$. The formation of galaxies crudely requires $\rho_{de} \leqslant 100\rho_{m0}$.

It is assumed that the primordial density fluctuation is about $\delta\rho/\rho \sim 10^{-5}$ in Weinberg's calculation. But in a full anthropic calculation, even this number is not to be presumed [10](for a complete discussion on "scanning parameters", see [11]). If we let $\delta\rho/\rho$ be a free parameter, we usually have $\rho_{de} \leqslant X\rho_{m0}$ and $X \gg 100!$, thus the currently observed value of dark energy can not be considered a consequence of the anthropic argument.

Weinberg also pointed that the age of the universe is a problem. He argues that if $\rho_{de}/\rho_m^0 = 9$ then $t_{age} = 1.1H_0^{-1}$. Actually, other people, including de Vauconleurs [12], Peebles [13], Turner, Steigman and Krauss [14] also considered this issue.

The anthropic argument is also applied to a possible negative cosmological constant. Let $\rho_{de} < 0$, then the Friedmann equation $3M_P^2 H^2 = \rho_m + \rho_{de}$ implies that when $\rho_m = |\rho_{de}|$, the universe starts to contract, thus $|\rho_{de}| \leqslant \rho_{m0}$.

5.3 Tuning Mechanism

The idea is to use a scalar to self-tune the stress tensor, which is a function of this scalar [15]. Assuming

$$\nabla^\mu \nabla_\mu \phi \sim T^\mu{}_\mu \sim \mathcal{R} \,, \qquad (5.11)$$

where $\mathcal{R}$ is the scalar curvature. If the trace of the stress tensor vanishes when ϕ rolls to a certain value ϕ_0, and ϕ will stay at this value ϕ_0 dynamically, then the effective cosmological constant becomes vanishing. However, it can be proven that

$$T_{\mu\nu} = e^{4\phi} g_{\mu\nu} \mathcal{L}_0 (\text{other fields}) \,, \qquad (5.12)$$

so this tuning mechanism can not be realized unless $\mathcal{L}_0 = 0$. But by requiring $\mathcal{L}_0 = 0$, the fine tuning of cosmological constant problem returns.

Another possibility is $\phi_0 = -\infty$. However in this case the effective Newtonian constant vanishes, because now ϕ_0 is a redefinition of G_N. This is again problematic.

5.4 Modifying Gravity

This is a rather popular theme now, but at the time when Weinberg wrote his review, there was only one proposal mentioned. This is the uni-modular metric theory [16, 17] (see [18, 19] for recent progress).

The idea is very simple. Removing the trace part from the Einstein equation

$$R_{\mu\nu} - \frac{1}{2} g_{\mu\nu} \mathcal{R} = 8\pi G T_{\mu\nu} \,, \qquad (5.13)$$

we will have

$$R_{\mu\nu} - \frac{1}{4} g_{\mu\nu} \mathcal{R} = 8\pi G (T_{\mu\nu} - \frac{1}{4} g_{\mu\nu} T) \,, \qquad (5.14)$$

where $T = T^\mu_\mu$. Assume $T_{\mu\nu}$ be conserved, namely $\nabla^\mu T_{\mu\nu} = 0$, we deduce from Eq. (5.14) that

$$\partial_\mu \mathcal{R} = -8\pi G \partial_\mu T \,, \qquad (5.15)$$

thus

$$8\pi G T = -R + 4\Lambda, \qquad (5.16)$$

where Λ is an integral constant, when we use Eq. (5.14) instead of Eq. (5.13) as a starting point. Substituting this back to the traceless Einstein equations, we obtain

$$R_{\mu\nu} - \frac{1}{2}g_{\mu\nu}\mathcal{R} = 8\pi G T_{\mu\nu} - \Lambda g_{\mu\nu}\,, \qquad (5.17)$$

so Λ is indeed the cosmological constant. That Λ is an integral constant is due to the fact that there are one fewer equations in the traceless Einstein equations. In fact Einstein himself considered this theory for a while.

The traceless equations can be considered as a consequence of the requirement that the determinant of the metric $g_{\mu\nu}$ is 1, so we have $g^{\mu\nu}\delta g_{\mu\nu} = 0$ or $\delta\sqrt{g} = 0$, we can introduce a Lagrangian multiplier into the action to enforce this.

As an integral constant, now Λ is considered as a free parameter in the theory to be determined by initial conditions, or to set a framework for utilizing the anthropic principle.

5.5   Quantum Cosmology

An accurate definition of quantum cosmology does not exist since there is no theory of quantum gravity to provide an appropriate framework.

The so-called quantum cosmology as advocated by Hawking can at most be considered as a qualitative picture of more accurate underlying theory. One starts with the Hamiltonian constraint

$$^{(3)}\mathcal{R} - 2\Lambda + N^{-2}(E_{ij}E^{ij} - E^2) - 2N^{-2}T_{00} = 0\,, \qquad (5.18)$$

where N is the lapse function as in $g_{00} = -N^2$, and $E^{ij} = \delta/\delta h_{ij}$, $^{(3)}\mathcal{R}$ is the scalar curvature of the three spatial geometry in the ADM splitting. The wave equation governing the wave function of the universe is called the Wheeler–DeWitt equation [20, 21],

$$\left(-G^{ij,lk}\frac{\delta^2}{\delta h^{ij}\delta h^{lk}} + {}^{(3)}\mathcal{R} - 2\Lambda - 2T_{00}\right)\Psi = 0\,, \qquad (5.19)$$

where Ψ is the wave function of the universe, it is a functional of the three geometry h_{ij} and other fields on the three dimensional spatial slice, however, it contains no time, and time has no place in the constraints. Actually, the Hamiltonian constraint is a consequence of the requirement of time-reparametrization invariance. As a consequence, we cannot impose the usual normalization condition on Ψ:

$$\int |\Psi(h_{ij}, \phi)|^2 \, [dhd\phi] = 1 \,. \tag{5.20}$$

Weinberg proposed to use one of the dynamical variables buried in h_{ij} or ϕ to replace the role of time.

There are infinitely many solutions to the Wheeler–DeWitt equation, if we assume this equation be well-defined. Hartle and Hawking proposed in [22] to select one out many by using the path integral

$$\Psi(h_{ij}, \phi) = \int [dgd\phi]_M e^{-S_E} \,, \tag{5.21}$$

where we assume that the three dimensional space Σ is the boundary of the four dimensional space M and g is an Euclidean metric on M, S_E is the Einstein–Hilbert action on this Euclidean four space obtained by an prescription of Wick rotation:

$$S_E = \frac{1}{16\pi G} \int \sqrt{g} (\mathcal{R} + 2\Lambda) + S_E(\phi) \,, \tag{5.22}$$

where $S_E(\phi)$ is the Euclidean action of the matter fields ϕ. Since the path integral is determined by the value of boundary metric h_{ij} and ϕ on Σ, there are no additional boundary conditions. This is called by Hartle and Hawking the no-boundary proposal of the wave function of the universe.

To have a probabilistic interpretation, consider a physical quantity A, a function of h_{ij} and ϕ on Σ. We define the probability of A assuming value A_0 be

$$P(A_0) = \int [dgd\phi]\delta(A(h, \phi) - A_0)e^{-S_E} \,. \tag{5.23}$$

However, Λ is a fixed parameter in action S_E. In order to compute the probability distribution of Λ, we need to make it variable. Hawking introduced a four form field strength in [23] to make the effective Λ a dynamic variable (see also [24]),

$$F_{\mu\nu\lambda\rho} = 4\partial_{[\mu} A_{\nu\lambda\rho]} \,, \tag{5.24}$$

with the action

$$S(A) = -\frac{1}{2 \times 4!} \int \sqrt{-g} F_{\mu\nu\lambda\rho} F^{\mu\nu\lambda\rho} \,. \tag{5.25}$$

The equation of motion for F is

$$\partial_\mu F^{\mu\nu\lambda\rho} = 0 \, . \tag{5.26}$$

Since F is totally asymmetric, let $F^{\mu\nu\lambda\rho} = F\epsilon^{\mu\nu\lambda\rho}/\sqrt{-g}$, the equation of motion leads to $\partial_\mu F = 0$, namely $F = \text{constant}$, and

$$S(A) = \frac{1}{2} \int \sqrt{-g} F^2 \, , \tag{5.27}$$

compared with the action

$$S(g, A) = \frac{1}{16\pi G} \int \sqrt{-g} (\mathcal{R} - 2\Lambda) \, , \tag{5.28}$$

we may conclude that $\Lambda \to \Lambda - 4\pi G F^2$. This is incorrect, since we need to consider the energy of this solution $E = \int d^3x \frac{1}{2} F^2$, and this leads to

$$\Lambda \to \Lambda + 4\pi G F^2 \, . \tag{5.29}$$

In the Euclidean action

$$S_{\mathrm{E}} = \frac{1}{16\pi G} \int (\mathcal{R} + 2\Lambda) \sqrt{g} + \frac{1}{2 \times 4!} \int \sqrt{g} F_{\mu\nu\lambda\rho} F^{\mu\nu\lambda\rho} \, , \tag{5.30}$$

let $F^{\mu\nu\lambda\rho} = F\epsilon^{\mu\nu\lambda\rho}/\sqrt{g}$, we have

$$S_{\mathrm{E}} = \frac{1}{16\pi G} \int (\mathcal{R} + 2(\Lambda + 4\pi G F^2)) \sqrt{g} \, , \tag{5.31}$$

this agrees with the consideration of energy.

Now, according to Eq. (5.23), we compute the probability distribution of Λ:

$$P(\Lambda_0) = \int [dg dA] \delta(\Lambda - \Lambda_0) e^{-S_{\mathrm{E}}} \, , \tag{5.32}$$

where $\Lambda = 4\pi G F^2$ (for simplicity we let the bare cosmological constant be zero). The most contribution to the above path integral comes from a classical solution to equation $3 M_{\mathrm{P}}^2 H^2 = \frac{1}{2} F^2$. We find

$$S_{\mathrm{E}} = -\frac{3\pi}{G\Lambda(F)} \, , \tag{5.33}$$

thus

$$P(\Lambda_0) \sim e^{\frac{3\pi}{G\Lambda_0}} . \tag{5.34}$$

Hawking concluded that $\Lambda_0 \to +0$ is the most probable value. But we now know that this is a wrong prediction!

Coleman later pointed out that Hawking's consideration is not the complete story. He suggested that one needs to take wormholes and baby universes into account [25]. Each type of baby universe is characterized by its physics properties, and we use i to label them. Let $a_i^\dagger$ be the creation operator for the baby universe of type i. Local creation of wormholes has the effect of modifying the action:

$$S \to \tilde{S} = S + \sum_i (a_i + a_i^\dagger) \int \mathcal{O}_i(x) , \tag{5.35}$$

this effective action works for the parent universe, and $\mathcal{O}_i$ is an local operator.

Consider a state without any wormhole, so $a_i|B\rangle = 0$ and

$$|B\rangle = \int \prod_i d\alpha_i f(\alpha_i)|\alpha\rangle,$$
$$(a_i + a_i^\dagger)|\alpha\rangle = \alpha_i|\alpha\rangle, \tag{5.36}$$

and

$$f(\alpha_i) = \prod_i \pi^{-1/4} e^{-\frac{\alpha_i^2}{2}} . \tag{5.37}$$

When acting on $|\alpha\rangle$, the action becomes

$$S \to S + \sum_i \alpha_i \int \mathcal{O}_i , \tag{5.38}$$

thus α_i becomes a coupling constant in the parent universe. The cosmological constant can be regarded as a coupling constant, its corresponding operator is the lowest dimensional operator $\mathcal{O}_1 = \sqrt{-g}$. In this framework, coupling constants including the cosmological constant become dynamical automatically. No additional mechanisms such as four form fluxes are needed. When observation is made on spacetime geometry, the observers will find the wave function of the universe to be in an eigenstate of $|\alpha\rangle$.

Coleman suggested that any apparently disconnect universes could be actually connected by wormholes. In this sense all the other apparently disconnected universes, which Hartle and Hawking ignore, should be summed over. This argument will not affect real constants in nature but does affect the effective "constants" which come from the baby universe creation operators (5.35). Using techniques well developed in quantum field theory (the summation of all vacuum to vacuum Feynman diagrams is the exponential of the summation of connected diagrams), one obtains

$$P(\alpha) = \exp\left(\int [dg dA] e^{-S_E}\right),\qquad(5.39)$$

and it leads to

$$P(\Lambda) \sim \exp\left(\exp(\frac{3\pi}{G\Lambda})\right),\qquad(5.40)$$

again this predicts $\Lambda = 0$, a wrong prediction.

References

[1] S. Weinberg, Rev. Mod. Phys. **61** (1989) 1.
[2] J. Wess and J. Bagger, *Supersymmetry and Supergravity* (Princeton University Press, Princeton, 1982).
[3] B. Zumino, Nucl. Phys. B **89** (1975) 535.
[4] E. Cremmer *et al.*, Phys. Lett. B **79** (1978) 231.
[5] J. R. Ellis *et al.*, Phys. Lett. B **134** (1984) 429.
[6] B. Carter, IAU Symp. **63** (1974) 291.
[7] R. H. Dicke, Nature **192** (1961) 440.
[8] R. H. Dicke, Phys. Rev. **125** (1962) 2163.
[9] S. Weinberg, Phys. Rev. Lett. **59** (1987) 2607.
[10] J. Garriga, A. Vilenkin, Prog. Theor. Phys. Suppl. **163** (2006) 245.
[11] S. Weinberg, in *Universe or Multiverse*, ed. Bernard Carr (Cambridge University Press, Cambridge, 2007) pp. 29–42.
[12] G. de Vaucouleurs, ApJ. **268** (1983) 468 ; Nature (London) **299** (1982) 303.
[13] P. J. E. Peebles, ApJ. **284** (1984) 439.
[14] M. S. Turner, G. Steigman, and L. M. Krauss, Phys. Rev. Lett. **52** (1984) 2090.
[15] A. D. Dolgov, in *Cambridge 1982, Proceedings, The Very Early Universe*, pp. 449–458.
[16] J. J. van der Bij, H. van Dam, and Y. J. Ng, Physica A **116** (1982) 307.
[17] W. G. Unruh, Phys. Rev. D **40** (1989) 1048.
[18] G. F. R. Ellis *et al.*, arXiv:1008.1196.
[19] T. Clifton *et al.*, arXiv:1106.2476.

[20] B. S. DeWitt, Phys. Rev. **160** (1967) 1113.
[21] R. Penrose, *Battelle Rencontres*, ed. C. DeWitt and J. A. Wheeler (Benjamin, New York, 1968).
[22] J. B. Hartle and S. W. Hawking, Phys. Rev. D **28** (1983) 1960.
[23] S. W. Hawking, Phys. Lett. B **134** (1984) 403.
[24] T. Banks, Nucl. Phys. B **249** (1985) 332.
[25] S. R. Coleman, Nucl. Phys. B **310** (1988) 643.

6
Symmetry

We have briefly reviewed Weinberg's classification, now we turn to more recent ideas and models about dark energy. We start from symmetry.

6.1 Supersymmetry in $2 + 1$ Dimensions

Witten pointed out in 1995 that supersymmetry in $2 + 1$ dimensions may help to solve the zero cosmological constant problem [1]. If such a theory exhibits the same phenomenon as the type IIA string theory in 10 dimensions, an additional dimension may emerge and becomes the third spatial dimension. The $2+1$ dimensional supersymmetry is smaller than the $3+1$ dimensional supersymmetry. As a result the unbroken $2+1$ dimensional supersymmetry (in $3+1$ spacetime with an emergent spatial dimension) does not contradict with observations and meanwhile still forces the cosmological constant be zero.

In $2 + 1$ dimensions, one can argue that there is no boson-fermion degeneracy. The existence of a particle of mass m creates a deficit angle in 2 spatial dimensions, $\theta \sim mG_3$, where G_3 is the 3 dimensional Newtonian constant. The appearance of the deficit angle makes the definition of charge impossible, thus the energy degeneracy becomes impossible too.

Becker, Becker and Strominger used an Abelian Higgs model to realize Witten's idea [2], but unfortunately there is no emergent third dimension in their model. Their model is a supergravity theory with field content $(\phi, A_\mu, N, \chi, \lambda) + (g_{\mu\nu}, \psi_\mu)$. BBS showed that the spectrum of vortices in this theory does not exhibit boson-fermion degeneracy.

6.2   't Hooft–Nobbenhuis Symmetry

't Hooft and later 't Hooft and Nobbenhuis proposed to consider symmetry transformation [3]

$$x^\mu \to iy^\mu, \quad p_x^\mu \to -ip_y^\mu. \tag{6.1}$$

As the simplest example, consider a scalar field

$$S = \int d^4x \left(-\frac{1}{2}(\partial\phi)^2 - V(\phi)\right), \tag{6.2}$$

with the stress tensor

$$T_{\mu\nu} = \partial_\mu\phi\partial_\nu\phi + g_{\mu\nu}\mathcal{L}(\phi). \tag{6.3}$$

Under transformation (6.1), or $x^\mu = iy^\mu$, $\partial_\mu^y = i\partial_\mu$,

$$\mathcal{L}_y = -\mathcal{L} = -\frac{1}{2}(\partial_\mu^y\phi)^2 + V, \\
T_{\mu\nu}^y = -T_{\mu\nu} = \partial_\mu^y\phi\partial_\nu^y\phi + g_{\mu\nu}\mathcal{L}. \tag{6.4}$$

We have in particular $T_{00}^y = -T_{00}$, however since $d^{D-1}x = i^{D-1}d^{D-1}y$, the Hamiltonian is not simply reversed in sign,

$$H^y = \int T_{00}^y d^{D-1}y = -(-i)^{D-1}H. \tag{6.5}$$

Let

$$\phi(x,t) = \int d^{D-1}p\left(a(p)e^{ipx} + a^\dagger(p)e^{-ipx}\right), \\
\pi(x,t) = \int d^{D-1}p\left(-ia(p)e^{ipx} + ia^\dagger(p)e^{-ipx}\right), \\
p^0 = (p^2 + m^2)^{\frac{1}{2}}, \tag{6.6}$$

we have

$$\phi(iy,it) = \int d^{D-1}q\left(a_y(q)e^{iqx} + a_y^\dagger(p)e^{-iqx}\right), \\
\pi_y = i\pi(iy,it) = \int d^{D-1}q\left(-ia_y(q)e^{iqx} + ia_y^\dagger(q)e^{-ipx}\right), \tag{6.7}$$

and

$$q^0 = (q^2 - m^2)^{\frac{1}{2}}, \qquad a_y(q) = (-i)^{D-1}a(p), \\
a_y^\dagger(q) = (-i)^{D-1}a^\dagger(p). \tag{6.8}$$

If $a^\dagger(p)$ is the Hermitian conjugate of $a(p)$, $a_y^\dagger$ is longer the Hermitian conjugate of a_y.

Note that if we demand $T_{\mu\nu} \to -T_{\mu\nu}$ be a symmetry, then $T_{00}|\Omega\rangle = 0$. Upon introducing gravity, let $g_{\mu\nu}^y = g_{\mu\nu}(x = iy)$ thus $ds_x^2 = -ds_y^2$, $R_{\mu\nu} \to -R_{\mu\nu}$, or $R_{\mu\nu}^y = -R_{\mu\nu}^x(iy)$. Start with the Einstein equation with a cosmological constant

$$R_{\mu\nu} - \frac{1}{2}g_{\mu\nu}\mathcal{R} + \Lambda g_{\mu\nu} = 8\pi G T_{\mu\nu}, \qquad (6.9)$$

we obtain

$$R_{\mu\nu}^y - \frac{1}{2}g_{\mu\nu}^y \mathcal{R}^y - \Lambda g_{\mu\nu}^y = 8\pi G T_{\mu\nu}^y, \qquad (6.10)$$

where we used $T_{\mu\nu} \to -T_{\mu\nu}$. Demanding $|\Omega\rangle$ be invariant, we deduce $\Lambda = 0$, This transformation maps a de Sitter space to an anti-de Sitter space.

't Hooft and Nobbenhuis pointed out that a scalar field and an Abelian gauge field can realize this symmetry transformation but

• The delta function $\delta^3(y)$ need be treated carefully in the second quantization scheme.

• $m^2 \to -m^2$, leading to tachyon.

• This symmetry can not be realized in a nonabelian gauge theory.

• The boundary conditions need be treated carefully. For instance, the boundary conditions at $x = \infty$ makes $H^y < 0$ in quantum mechanics, and makes it equal to $-iH^x$ in quantum field theory.

This concludes our discussion on the 't Hooft–Nobbenhuis symmetry.

6.3 Kaplan–Sundrum Symmetry

This symmetry is quite similar to the 't Hooft–Nobbenhuis symmetry. Kaplan and Sundrum [4] proposed that to each matter field ψ there is a ghost companion $\tilde{\psi}$, the Lagrangian is

$$\mathcal{L} = \sqrt{-g}\left(\frac{M_p^2}{2}\mathcal{R} - \Lambda\right) + \mathcal{L}_{\text{matter}}(\psi, D_\mu) - \mathcal{L}_{\text{matter}}(\tilde{\psi}, D_\mu), \qquad (6.11)$$

where the form of $\mathcal{L}_{\text{matter}}(\psi)$ and the form of $\mathcal{L}_{\text{matter}}(\tilde{\psi})$ are identical. We see that the name ghost is appropriate since the kinetic term of $\tilde{\psi}$ has a wrong sign.

Ignoring gravity for a while, there is a transformation between ψ, $\tilde{\psi}$:

$$P : \psi \to \tilde{\psi} , \qquad \tilde{\psi} \to \psi . \tag{6.12}$$

Under this, $H \to -H$, since $H = H(\psi) - H(\tilde{\psi})$. Namely $PH = -HP$ or $\{P, H\} = 0$. For a vacuum $|0\rangle$, $P|0\rangle = |0\rangle$, then

$$\langle 0|\{P, H\}|0\rangle = 2\langle 0|H|0\rangle = 0 , \tag{6.13}$$

thus if $|0\rangle$ is an eigenstate of H, $H|0\rangle = 0$.

When gravity is introduced, this antisymmetry is broken. Let $P : g_{\mu\nu} \to g_{\mu\nu}$, the Hamiltonian

$$\mathcal{H} = -G^{ij,lk} \frac{\delta^2}{\delta h_{ij} \delta h_{lk}} - 2\Lambda - 2T_{00} +^{(3)} \mathcal{R} \tag{6.14}$$

does not anticommute with P. If the wave function Ψ has $P\Psi = \Psi$, then

$$\mathcal{H} P \Psi = \left(-G^{ij,lk} \frac{\delta^2}{\delta h_{ij} \delta h_{lk}} +^{(3)} \mathcal{R} - 2\Lambda - 2T_{00} \right) P\Psi$$

$$= \left(-G^{ij,lk} \frac{\delta^2}{\delta h_{ij} \delta h_{lk}} +^{(3)} \mathcal{R} - 2\Lambda + 2T_{00} \right) \Psi , \tag{6.15}$$

we deduce $T_{00} \Psi = 0$.

As an example, consider a scalar ϕ, then the ghost companion is $\tilde{\phi}$,

$$\mathcal{L} = \frac{1}{2}(\partial_\mu \phi)^2 - \frac{1}{2}m^2\phi^2 - \lambda\phi^4 - \frac{1}{2}(\partial_\mu \tilde{\phi})^2 + \frac{1}{2}m^2\tilde{\phi}^2 + \lambda\tilde{\phi}^4 . \tag{6.16}$$

Fixing $g_{\mu\nu}$, the path integral in a quantum theory is

$$\int [d\phi d\tilde{\phi}] e^{iS(\phi) - iS(\tilde{\phi})} . \tag{6.17}$$

The propagator of ϕ is forward with positive energy, thus $i\epsilon$ prescription is used, while the propagator of $\tilde{\phi}$ is backward for positive energy, $-i\epsilon$ prescription is used. We then infer $S_{\text{eff}}(\phi, \tilde{\phi}) = S_{\text{eff}}(\phi) - S_{\text{eff}}(\tilde{\phi})$. In particular, if there is a term $\Lambda \int \sqrt{-g}$ in $S_{\text{eff}}(\phi)$, this term is canceled by a term in $S_{\text{eff}}(\tilde{\phi})$.

Now consider the effect of quantum gravity. The interesting aspect of this symmetry is that the effect of gravity will introduce a small Λ. Since there is no ghost companion of $g_{\mu\nu}$, the quantum fluctuation of the metric introduces a term $\Lambda \int \sqrt{-g}$. Let the cutoff be μ, we must have

$$\Lambda \sim \mu^4 . \tag{6.18}$$

Since $\Lambda \sim (2 \times 10^{-3} \text{ eV})^4$, $\mu \leqslant 2 \times 10^{-3}$ eV, or $\mu^{-1} \geqslant 30$ microns. We know that Newtonian gravity is tested above this scale, so it is possible that quantum gravity may break Newtonian gravity below this scale.

We see that the local quantum effects of ψ and $\tilde{\psi}$ cancel exactly, while non-local effects may not cancel, so it is possible to have a contribution to Λ: $\Lambda \to \Lambda + \mu^6/M_P^2$, this is much smaller than μ^4.

If ψ and $\tilde{\psi}$ are coupled, the vacuum is unstable. For example, for $g^2\phi^2\tilde{\phi}^2$, there is a process $|0\rangle \to \phi + \phi + \tilde{\phi} + \tilde{\phi}$, the amplitude is divergent:

$$P_{0\to\phi^2\tilde{\phi}^2} = g^2 \int \prod d^4 p_i d^4 k_i \prod \delta(p_i^2 - m^2)\delta(k_i^2 - m^2)\delta^4\left(\sum(p_i + k_i)\right)$$

$$= \infty . \tag{6.19}$$

Let $s = (p_1 + p_2)^2$, the integral in the above amplitude is divergent, we need to introduce a cutoff for s, $s_{\max}$, and a cutoff for $p_1^0 \leqslant \epsilon$, then

$$P_{0\to\phi^2\tilde{\phi}^2} \sim g^2\epsilon^2 s_{\max} . \tag{6.20}$$

Even without direct coupling between a field and a ghost field, their coupling to the metric also triggers instability. For example, we estimate

$$P_{0\to\gamma\gamma\tilde{\gamma}\tilde{\gamma}} \sim \frac{1}{4\pi}\left(\frac{1}{8\pi}\right)^2 \frac{\mu^8}{M_P^4}$$

$$\sim 2 \times 10^{-92} \left(\frac{\mu}{2 \times 10^{-3} \text{ eV}}\right)^8 (\text{cm}^3 \times 10 \text{ Gyr})^{-1} . \tag{6.21}$$

This term is negligible.

It is interesting to notice that the force between a field and its ghost companion is repulsive. Next, we consider the effect of potential of scalar fields. Suppose there are two local minima in $V(\psi)$, then there are two local maxima in $-V(\tilde{\psi})$, if ψ runs to the global minimum, $\tilde{\psi}$ runs to the global maximum, the values cancel exactly.

6.4   Symmetry of Reversing Sign of the Metric

Recai Erdem considered this kind of symmetry [5]. Consider action in D spatial dimensions

$$S(g) = \frac{1}{16\pi G} \int \sqrt{(-1)^D g}\,\mathcal{R} , \tag{6.22}$$

this is a generalization of the Einstein–Hilbert action. We demand this action be invariant under the reflection of the metric $g_{AB} \to -g_{AB}$. Since $\mathcal{R} \to -\mathcal{R}$, we have $\sqrt{(-1)^D g}\mathcal{R} \to -\sqrt{(-1)^{D+D+1}g}\mathcal{R}$. Namely

$$(-1)^{D+\frac{1}{2}+1} = (-1)^{D/2}, \qquad (6.23)$$

or $(-1)^{(D+1)/2} = -1$, we deduce $D + 1 = 2(2n + 1)$. When $n = 0$, the dimensionality of spacetime is 2, but $D + 1 = 4$ is not a solution.

Since $(-1)^{(D+1)/2} = -1$, we know that $\sqrt{-g}$ changes sign, and the cosmological constant term is not invariant, thus forbidden by this symmetry.

This symmetry is preserved in the action of a scalar field

$$-\int d^{D+1}x\sqrt{-g}\frac{1}{2}g^{AB}\partial_A\phi\partial_B\phi. \qquad (6.24)$$

The action of a fermionic field is not invariant, unless we demand $\gamma^A \to -\gamma^A$. The worst thing is that the stress tensor $T_{AB} = \partial_A\phi\partial_B\phi - \frac{1}{2}g_{AB}\mathcal{L}$ has an invariant part so the vacuum expectation of T_{00} is not vanishing.

6.5   Scaling Invariance in $D > 4$

There may be many such approaches, a typical one was proposed by Wetterich [6]. He postulates a dilatation field ξ, when we rescale the metric $g_{AB} \to \lambda^{-2}g_{AB}$, $\chi \to \lambda^{(D-3)/2}\xi$, the action

$$S(\xi) = \int \sqrt{-g}\left(-\frac{1}{2}\xi^2\mathcal{R} + \frac{\alpha}{2}\partial^\mu\xi\partial_\mu\xi\right) \qquad (6.25)$$

is invariant. If there is a potential $V(\xi)$, we require $\lambda^{-D}V(\lambda^{(D-2)/2}\xi) = V(\xi)$. If $V(\xi) = \xi^\nu$, then $\nu = 2D/(D-2)$. When $D = 4, 6$, ν is an integer. If $D > 6$, $V = 0$.

Let Ω be the internal volume, $\Omega = \int d^{D-4}x\sqrt{h}$, let $\chi = \Omega^{\frac{1}{2}}\xi$, the effective four dimensional action is

$$\Gamma = \frac{1}{2}\int \sqrt{-g}\chi^2\mathcal{R} + \frac{\alpha}{2}\int \sqrt{-g}(\partial\chi)^2. \qquad (6.26)$$

There is no term $\lambda\chi^4$, and $\chi = \chi_0$ is a solution. However, there is a problem of stability.

References

[1] E. Witten, Mod. Phys. Lett. A **10** (1995) 2153.

[2] K. Becker, M. Becker, and A. Strominger, Phys. Rev. D **51** (1995) 6603.

[3] G. 't Hooft and S. Nobbenhuis, Class. Quant. Grav. **23** (2006) 3819.

[4] D. E. Kaplan and R. Sundrum, JHEP **0607** (2006) 042.

[5] R. Erdem, Phys. Lett. B **621** (2005) 11.

[6] C. Wetterich, Phys. Rev. Lett. **102** (2009) 141303; C. Wetterich, Phys. Rev. D **81** (2010) 103507.

7
Anthropic Principle

If the anthropic principle is the reason for $\Lambda \sim 0$, one of the necessary conditions is that, Λ is either a continuous variable, or if it is discrete, the interval $\Delta\Lambda$ must be sufficiently small.

Bousso and Polchinski are probably the first to point out that the second possibility is realized in string theory [1] (see also [2] for a review). This is the beginning of studies on the string landscape.

7.1 Bousso–Polchinski Scenario

There are many totally anti-symmetric 3-form fields in string theory, for instance, there is a membrane coupled 3-form $A_{\mu\nu\rho}$ in M-theory. In the type IIA string theory, the field $C^{(3)}_{\mu\nu\rho}$ is coupled to D2-branes. When IIA (B) theory is compactified on a Calabi–Yau manifold CY_3, other C fields induce a number of 3-form fields. Let Σ_{p-2} be a $p-2$ circle in CY_3, we have

$$\int_{\Sigma_{p-2}} C^{p+1} \rightarrow C_{\mu\nu\rho} \,, \tag{7.1}$$

different p and different Σ_{p-2} give rise to a different 3-form field in 4 dimensions.

For instance, in IIA string theory, $C^{(5)}$ on Σ_2, $C^{(7)}$ on Σ_4; in IIB string theory, $C^{(6)}$ on Σ_3, but there is no Σ_5 or Σ_1, so $C^{(4)}$ and $C^{(8)}$ do not introduce 3-form fields.

Let us focus on a single 3-form field $C_{\mu\nu\rho}$, its four form strength is $F_4 = 4dC$ with an action

$$S = \int \sqrt{-g} \left(\frac{1}{2\kappa^2} (\mathcal{R} - 2\Lambda_0) - \frac{Z}{2 \times 4!} F_4^2 \right) + S_{\text{brane}} \,, \tag{7.2}$$

where Z is a normalization constant. The equation of motion

$$\partial_\mu(\sqrt{-g}F^{\mu\nu\rho\sigma}) = 0\,, \tag{7.3}$$

has a solution

$$F^{\mu\nu\rho\sigma} = C\epsilon^{\mu\nu\rho\sigma}/\sqrt{-g}\,, \tag{7.4}$$

this leads to

$$\rho_{\text{de}} = \Lambda = \Lambda_0 + \frac{Z}{2}C^2\,. \tag{7.5}$$

It may not be obvious that the constant C is quantized. Indeed if spacetime is really four dimensional, C is a continuous parameter. If spacetime is higher dimensional, C is quantized. For example, in the 11 dimensional M theory, we have

$$S = 2\pi M_{11}^9 \int \sqrt{-g}\left(\mathcal{R} - \frac{1}{2\times 4!}F_4^2\right)\,, \tag{7.6}$$

where M_{11} is the eleven-dimensional Planck mass. A M5-brane is coupled to A_6, the dual of A_3. The coupling is

$$2\pi M_{11}^6 \int A_6\,, \tag{7.7}$$

this coupling leads to

$$2\pi M_{11}^6 \int_{M_7} F_7 = 2\pi n\,, \tag{7.8}$$

this is the standard Dirac quantization condition. Now, F_7 is dual to F_4,

$$F_7^{\mu_1\cdots\mu_7} = \frac{1}{4!}\epsilon^{\mu_1\cdots\mu_7\nu_1\cdots\nu_4}F_{\nu_1\cdots\nu_4}\,. \tag{7.9}$$

Let $\mu_1\cdots\mu_7 \in M_7$, then $F_7 = \epsilon^{\mu_1\cdots\mu_7}F_0$, and $2\pi M_{11}^6 V_7 F_0 = 2\pi n$, thus F_0 is quantized as $F_0 = n/(M_{11}^6 V_7)$. The reduced action is

$$S = 2\pi M_{11}^9 V_7 \int \sqrt{-g}\left(\mathcal{R} - \frac{1}{2\times 4!}F_4^2\right)\,, \tag{7.10}$$

thus

$$\frac{1}{2\kappa^2} = 2\pi M_{11}^9 V_7 = Z\,, \qquad F_0 = \frac{n}{M_{11}^6 V_7} = \frac{2\pi n M_{11}^3}{Z}\,. \tag{7.11}$$

For a M2-brane, the charge is $e = 2\pi M_{11}^3$, namely the coupling is $e \int A_3 = 2\pi M_{11}^3 \int A_3$. We have $F_0 = ne/Z$. Now, for solution $F^{\mu_1 \cdots \mu_4} = C\epsilon^{\mu_1 \cdots \mu_4}/\sqrt{-g}$, we have

$$F_0 = \frac{1}{4!}\epsilon_{\mu_1 \cdots \mu_4} F^{\mu_1 \cdots \mu_4} = C = \frac{ne}{Z}, \qquad (7.12)$$

we see that C is quantized. Our conclusion is that, if the action is

$$S(A) = -\frac{Z}{2 \times 4!} \int F_4^2, \qquad (7.13)$$

and the charge coupling is $e \int A_3$, then $C = ne/Z$.

In formula

$$\rho_{\mathrm{de}} = \Lambda_0 + \frac{Z}{2}C^2 = \Lambda_0 + \frac{n^2 e^2}{2Z}, \qquad (7.14)$$

the dimensions of the constants are $[e] = M^3$, $[Z] = M^2$. If $e^2 \sim M_P^6$, $Z \sim M_P^6 L^7$, then $\Delta\Lambda \geqslant M_P^{-3}L^{-7}$. Let $M_P^{-3}L^{-7} \sim (2 \times 10^{-3} \text{ eV})^4$, then $L^{-7} \sim 10^7 \text{ GeV}^7$, or $L^{-1} \sim 10 \text{ GeV}$. This is possible, but there is a problem: suppose the bare cosmological constant $\Lambda_0 \sim -M_P^4$ thus $n^2 \sim M_P^4 Z/e^2$, so $\Delta\Lambda \propto ne^2/Z \sim M_P^2 e/\sqrt{Z}$. Use $e \sim M_P^3$, $\sqrt{Z} \sim M_P^{9/2}L^{7/2}$ we infer $\Delta\Lambda \sim \sqrt{M_P}L^{-7/2} \sim (2 \times 10^{-12} \text{ GeV})^4$, $L^{-7} \sim 10^{-64} \text{ GeV}^7$, or $L^{-1} \sim 10^{-9} \text{ GeV} = 1 \text{ eV}$, this is too small, or L is too large.

To solve the above problem, Bousso and Polchinski proposed to consider multiple 3-form fields $C_{\mu\nu\lambda}^a$, $a = 1, \cdots, J$. Let $Z_i = 1$ and $e \to q_i$, then

$$\Lambda = \Lambda_0 + \frac{1}{2}\sum_{i=1}^{J} n_i^2 q_i^2. \qquad (7.15)$$

Thus, the additional term is a distance squared in the n-dimensional Euclidean space with fundamental lattice spacing q_i. The volume of a fundamental cell is

$$\Delta V = \prod_{i=1}^{J} q_i. \qquad (7.16)$$

Now, consider a shell in between r and $r + \Delta r$, the volume of this shell is

$$\Omega_J r^{J-1} \Delta r. \qquad (7.17)$$

If this volume is greater than ΔV, then there is at least one lattice point falling into this shell.

Let $\Lambda_0 = -r^2/2$, and

$$\Lambda = \frac{1}{2}\left(\sum_{i=1}^{J} n_i^2 q_i^2 - r^2\right) = \Delta\Lambda. \tag{7.18}$$

Let $\{n_i\}$ fall in between r and $r + \Delta r$, there must be $r\Delta r = \Delta\Lambda$, $\Delta r = r^{-1}\Delta\Lambda$, then

$$\Omega_J r^{J-1}\Delta r = \Omega_J r^{J-2}\Delta\Lambda = \Omega_J (2|\Lambda_0|)^{(J-2)/2}\Delta\Lambda. \tag{7.19}$$

This volume must be no smaller than the volume of a fundamental cell, so

$$\prod_{i=1}^{J} q_i \leqslant \Omega_J(2|\Lambda_0|)^{(J-2)/2}\Delta\Lambda. \tag{7.20}$$

The physical Λ is equal to $\Delta\Lambda$, the smallest allowed value is

$$\Lambda = \frac{\displaystyle\prod_{i=1}^{J} q_i}{\Omega_J(2|\Lambda_0|)^{J/2-1}}, \tag{7.21}$$

we see that if $q_i < \sqrt{|\Lambda_0|}$, it is easy to have a very small Λ. Ω_J, the volume of the unit sphere in J-dimensional Euclidean space, is $2\pi^{J/2}/\Gamma(J/2)$. Let $q_i = 100^{-1}\sqrt{|\Lambda_0|}$, we find that when $J \sim 100$, we have roughly

$$\frac{\Lambda}{|\Lambda_0|} \sim 10^{-120}. \tag{7.22}$$

This is how the Bousso–Polchinski scenario solves the cosmological constant problem.

Just how to generate the right quantum numbers n_i? There is the Brown–Teitelboim mechanism [3, 4] to use. The BT mechanism is similar to electron-positron creation in an electric field in $1 + 1$ dimensions.

In $1 + 1$ dimensions , let there be a pair of charges q and $-q$, $\partial_x E = -q\delta(x) + q\delta(x + L)$, the solution is

$$E(x) = \begin{cases} E, & x < 0; \\ E - q, & 0 < x < L; \\ E, & X > L. \end{cases} \tag{7.23}$$

So before the pair creation, the electric field strength is E everywhere, and becomes $E - q$ in between the two charges after creation. It decreases until

E reaches the minimum $E_{\min} = E - [E/q]q \leqslant q$. The rate of pair creation is

$$P \propto \exp\left(-\frac{c}{q^2 E^2}\right). \tag{7.24}$$

Similarly, a spherical membrane is created in 4 dimensions in a background of F_4 and $F_4 \to F_4 - C = F_4 - q$, and the cosmological constant is shifted as

$$\Lambda \to \Lambda + \frac{1}{2}(F_4 \pm q)^2 - \frac{1}{2}F_4^2. \tag{7.25}$$

But this process is very slow and in the end we need the help of the anthropic principle. The difference between membrane creation and pair creation is that the membrane creation can either decrease F_4 or increase it, however the probability of decreasing Λ is greater than that of increasing Λ.

In the original BT model, there is an empty universe problem. This is because in the original BT, there is only one kind of membrane, thus q is required to be very small in order to have a very small Λ. But before the membrane creation $\Lambda - \Delta\Lambda$ is also very small, thus the decay rate of this previous universe is extremely small such that inflation in this phase makes the universe almost empty.

This problem is eliminated in the BP model, since before the last transition, the previous Λ is large and the inflation period is short. The major problem remains in the BP model is the moduli stabilization.

7.2 KKLT Scenario

In the Bousso–Polchinski scenario, charges q_i as well as normalization constants Z_i are all moduli dependent. To solve this problem, Giddings, Kachru and Polchinski proposed to introduce D7-branes as well as fluxes of $H^{(3)}$ and $F^{(3)}$ in compactification [5].

A D7-brane is specified by a complex function τ over CY_3 [6], $H^{(3)}$ and $F^{(3)}$ form a complex 3-form $G^{(3)} = F^{(3)} - \tau H^{(3)}$. Let

$$\tilde{F}^{(5)} = F^{(5)} - \frac{1}{2}C^{(2)} \wedge H^{(3)} + \frac{1}{2}B \wedge F^{(3)} = (1 + *)[d\alpha \wedge dx^0 \wedge \cdots \wedge dx^3], \tag{7.26}$$

then

$$d\Lambda + \frac{i}{\operatorname{Im}\tau}d\tau \wedge Re\Lambda = 0, \qquad \Lambda = e^{4A}_{*(6)}G^{(3)} - i\alpha G^{(6)}, \tag{7.27}$$

where $*(6)$ is the dual operation in the 6 manifold CY_3, A is the warp factor in the warped metric

$$ds^2 = e^{2A(y)}\eta_{\mu\nu}dx^\mu dx^\nu + e^{-2A(y)}ds^2_{CY}, \qquad (7.28)$$

and y are coordinates on CY_3. This helps to fix many of the moduli except for all the Kähler moduli, for example, the complex scalar ρ associated with the scale of the Calabi–Yau CY_3. This is the scalar appearing in a no-scale supergravity theory.

The warp factor in the GKP model offers us means to solve the hierarchy problem, as in the Randall–Sundrum scenario [7].

Consider a complex structure moduli, corresponding to a conifold (z is the size of S^3 in $S^2 \times S^3$), let

$$\frac{1}{2\pi\alpha'}\int_{A=S^3} F^{(3)} = 2\pi M, \qquad \frac{1}{2\pi\alpha'}\int_B H^{(3)} = -2\pi K, \qquad (7.29)$$

then z is stabilized to

$$z \sim e^{-\frac{2\pi K}{M g_s}}, \qquad e^{A_{\min}} = e^{-\frac{2\pi K}{3M g_s}}. \qquad (7.30)$$

KKLT first considered a single Kähler moduli [8], and pointed out that there are two effects to help to fix the radial Kähler moduli.

(a) If there is a four dimensional complex sub-manifold in CY_3, then wrapping the Euclidean D3-brane on this sub-manifold forms D-instantons. The contribution to the superpotential has a form

$$W \sim \exp(2\pi i\rho). \qquad (7.31)$$

(b) Fluxes induce D7-branes, and the field theory on D7-branes is $N = 1$ super Yang–Mills theory, and we have

$$\frac{4\pi}{g^2_{YM}} = \text{Im}\rho. \qquad (7.32)$$

There is a superpotential

$$W = A\exp\left(\frac{2\pi i\rho}{N_c}\right). \qquad (7.33)$$

Let

$$W = W_0 + Ae^{2\pi ia\rho}, \qquad K = -2\ln(-i(\rho - \bar\rho)), \qquad (7.34)$$

then $DW = 0$ leads to

$$W_0 = -Ae^{-a\sigma_c}\left(1 + \frac{2}{3}a\sigma_c\right),\tag{7.35}$$

where $\sigma_c = \text{Im}\rho$, and we get

$$V = -a^2 A^2 e^{-2a\sigma_c}/(6\sigma_c) < 0\,,\tag{7.36}$$

thus the cosmological constant is negative and the universe is an anti-de Sitter space AdS_4, supersymmetry is unbroken.

If fluxes are not balanced, we need to introduce anti-branes, namely $\overline{\text{D3}}$-branes. The existence of anti-branes breaks supersymmetry, and all moduli will be fixed. Let $a_0 = \exp(2A_0)$ be the warp factor, then $\overline{\text{D3}}$ contribute to the energy density by a factor

$$\Delta V = \frac{2a_0^4 T_3}{g_s^4}\frac{1}{(\text{Im}\rho)^3}\,,\tag{7.37}$$

where T_3 is the D3-brane tension. Thus, $V \to V + \Delta V$. Although ΔV is small, but $V + \Delta V > 0$, the anti-de Sitter space is modified to become a de Sitter space.

Next, KKLT argued that although the de Sitter space is metastable, the life time is much longer than the age of our universe, but smaller than the Poincare recurrence time $t \sim \exp(S_0) \sim \exp(10^{120})$. All the stable and metastable anti-de Sitter spaces and de Sitter spaces form the string landscape.

There have been a lot of efforts invested to study statistics of the string landscape. There are in principle infinitely many metastable vacua of the Bousso–Polchinski type, but many of them are not reliable. In the KKLT models, are there upper limits for the flux numbers of $F^{(3)}$ and $H^{(3)}$? Someone estimated that there are at least [9]

$$10^{500}\tag{7.38}$$

metastable vacua.

Susskind [10] invented terminology "string landscape", and argued that our universe may be multiverse consisting numerous regions in which physics varies. The implication of the multiverse is to be discussed in the following section.

7.3 Populating the Landscape and Anthropic Interpretations

The vast string landscape itself does not lead to an anthropic interpretation for cosmological constant. Instead, the landscape must be populated.[1] In other words, one needs a mechanism to produce different universes in a (at best connected patch of) multiverse.

There are different approaches to populate the landscape. For example, Hawking and Hertog [12] pointed out that the wave function of universe is one populating method. Different observers live in their different histories, and they are summed over in the no boundary path integral. In other words, observers in different universes live in different decohered branches of a single wave function. On the other hand, here we shall mainly discuss another better studied scenario: eternal inflation.

When the quantum fluctuation of the inflaton $\delta\phi = \frac{H}{2\pi}$ is larger that the rolling of ϕ in a Hubble time, namely $\delta\phi = \frac{H}{2\pi} > \Delta\phi = \dot{\phi}H^{-1}$, eternal inflation occurs [13, 14]. Whether eternal inflation really happens is a matter of controversy [15–18]. For instance, the weak gravity conjecture [19] may prohibit it to occur [17].

On the other hand, de Sitter space itself may not be eternal [20, 21], it has a finite life time, for instance its life time can not be longer than the Poincare recurrence time if we view this spacetime has finite dimension of the Hilbert space. It was argued in [20, 21] that a universe with conditions all the same as our universe except the CMB temperature is higher is more likely, the probability of its occurrence is $\sim e^{S_i - S}$ where S = entropy of the pure de Sitter space, S_i = entropy of a particular universe. It is also pointed out that a de Sitter space is a resonant state in the multiverse.

Anyway, eternal inflation is still a possibility which is semi-classically well defined. Before a more complete quantum theory of gravity clarifies all the subtleties, we have to take eternal inflation seriously. If eternal inflation indeed happens, our universe is in a small part of the eternal inflating universe. The situation is like our earth is a small part of our observable universe. This provides a playground for the anthropic principle.

The validity of the anthropic principle is very controversial, there are a number of problems need to be addressed, including:

- The measure of the multiverse.

---

[1] Except that, if one (much more aggressively) assumes every self-consistent mathematical structure is automatically "populated" by a higher level of the multiverse, see [11].

It is intuitive to imagine that a "typical" vacuum in the landscape is a kind of vacuum that is "realized" in the multiverse most frequently. However, it is rather difficult to realize this idea. A number of measures have been proposed in the literature, including the volume based measure [22] (see [23] with scale factor cutoff), the local measure [24, 25], the Liouville measure [26–28] the stochastic measure [29], and so on. However, there is so far no principle to determine which measure to use. On the other hand, the measure problem is also closely related with other problems, as discussed below.

• How to correctly apply Bayesian statistics?

The measure problem addresses the *a priori* probability from the theoretical point of view. After that the anthropic probability must be assigned to have anthropic predictions. The anthropic probability is proportional to the product of the *a priori* probability and a conditional probability for the theory to be observed, according to the Bayesian statistics:

$$P(\text{theory } x|\text{selection}) = \frac{P(\text{selection}|\text{theory } x)P(\text{theory } x)}{\sum_{y} P(\text{selection}|\text{theory } y)P(\text{theory } x)} . \qquad (7.39)$$

The "selection" here can be understood as an anthropic effect. Unfortunately, there is no principle stating how to properly define the probability from anthropic selection effect. For example, Page [15, 30, 31] assumed typical observers to make observations, Hartle and Srednicki [32] argued it is exactly us human observers instead of a typical observer that should be used here. On the other hand, some of us [33] argued a compromise treatment seems more natural, that the anthropic probability should be assigned by the probability of existence, instead of the number of observers. Alternatively, Bousso, Harnik, Kribs and Perez [34] also proposed a more operable entropic measure to mimic the anthropic measure.

• The problem of Boltzmann brains.

In the concluding part of [35], Boltzmann described the idea of his assistant Schuetz: we may come from thermal fluctuations in an extremely large universe, whereas the whole universe is in a thermal equilibrium state. Following this idea, one eventually arrives at a paradox: the so-called Boltzmann problem.

Now we are not going to talk about whether our galaxies, stars and planets originate from thermal fluctuations or not. Instead, we refer ourselves as "human observers", who live in a low entropy environment and use entropy increasing to maintain their lives. However, if the universe is large enough and in a thermal state (indeed in an asymptotically de Sitter universe with Gibbons–Hawking temperature), there will also be other observers originated completely from thermal fluctuations, who are by pure chance be in a low entropy state themselves, and isolated from the thermal equilibrium environment. These observers are called Boltzmann brains, or freak observers.

In principle, we can not assert we are not Boltzmann brains but by the next moment we will be almost sure, by observing that ourselves are not returning to the thermal equilibrium. If we were Boltzmann brains, this probability is exponentially small. However, why are we not Boltzmann brains?

As an example, if our universe starts from a big bang and end up asymptotically de Sitter, in a finite comoving volume the number of human observers is finite, because the available matter entropy difference will eventually be used up.[2] However, the number of Boltzmann brains is infinite, as long as de Sitter space does not decay. If we were typical observers, we should have inferred that we were Boltzmann brains, not human observers. This contradicts with the observation in the last paragraph, that we are not Boltzmann brains. Similar paradox happens in an eternally inflating universe.

Even worse, Bousso, Freivogel, Leichenauer and Rosenhaus [36] argued that time will end for eternal inflation with the assumption of typicality. This is too bad since no observer is even worse than too many observers.

• Is there any definite prediction of the anthropic principle?

It is in debate whether anthropic principle could make predictions. Some may say that Weinberg has already predicted the cosmological constant from anthropic principle and others may argue that what Weinberg actually did is that he cannot calculate cosmological constant from first principle, and simply assumes a probable value from existence of human.

More generally, in principle one can construct a measure for the whole landscape, and predict where is most probable for human beings. If this prediction agrees with current experiments, we may take the view point that anthropic principle indeed make predictions. But again, the answer depends on the definition of prediction, because the prediction that anthropic principle does is logically different from predictions of traditional science.

---

[2] To be more precise, for a human observer, one at least must be able to remember something. To prepare the memory to remember anything, there has to be entropy increasing. When entropy is maximized, there is no available entropy difference to remember anything.

To understand anthropic principle, it is good to return to the much better understood question why we live on the earth, with an environment surprisingly suitable for human beings. However, even in this much easier case, we can not yet make predictions or answer whether we live on a most typical planet suitable for intelligence. The way for anthropic principle to make predictions for fundamental physics, is thus much longer.

• There is evidence that there are just too many metastable vacua (Witten: M2-branes $\rightarrow SU(n) \times SU(n)/(\text{discrete symmetry}) \rightarrow C^m_{m+n-1} \times C^n_{m+n-1}$ vacua. Take $m = n \sim 10^{100}$, then the number of vacua of this type is $(C^n_{2n})^2 \sim 2^{4n} \sim 10^{10^{100}}$, a googolplex!)

Although there are many problems with the anthropic principle, we can not yet rule it out.

References

[1] R. Bousso and J. Polchinski, JHEP **0006** (2000) 006.
[2] J. M. Cline, arXiv:hep-th/0612129.
[3] J. D. Brown and C. Teitelboim, Phys. Lett. B **195** (1987) 177.
[4] J. D. Brown and C. Teitelboim, Nucl. Phys. B **297** (1988) 787.
[5] S. B. Giddings, S. Kachru, and J. Polchinski, Phys. Rev. D **66** (2002) 106006.
[6] C. Vafa, Nucl. Phys. B **469** (1996) 403.
[7] L. Randall and R. Sundrum, Phys. Rev. Lett. **83** (1999) 3370; Phys. Rev. Lett. **83** (1999) 4690.
[8] S. Kachru *et al.*, Phys. Rev. D **68** (2003) 046005.
[9] M. R. Douglas, JHEP **0305** (2003) 046.
[10] L. Susskind, In *Universe or Multiverse*, ed. Bernard Carr (Cambridge University Press, Cambridge, 2007) pp. 247.
[11] M. Tegmark, Sci. Am. **288** (2003) 30.
[12] S. W. Hawking and T. Hertog, Phys. Rev. D **73** (2006) 123527.
[13] P. J. Steinhardt, *The Very Early Universe*, in *Proceedings of the Nuffield Workshop*, ed. G. Gibbons, S. W. Hawking, S. T. C. Siklos (Cambridge University Press, 1982).
[14] A. Vilenkin, Phys. Rev. D **27** (1983) 2848.
[15] D. N. Page, Phys. Lett. B **669** (2008) 197.
[16] N. Arkani-Hamed *et al.*, JHEP **0803** (2008) 075.
[17] Q. G. Huang, M. Li, and Y. Wang, JCAP **0709** (2007) 013.
[18] Y. Wang, arXiv:0805.4520.
[19] N. Arkani-Hamed *et al.*, arXiv:hep-th/0601001.
[20] L. Dyson, M. Kleban, and L. Susskind, JHEP **0210** (2002) 011.
[21] N. Goheer, M. Kleban, and L. Susskind, JHEP **0307** (2003) 056.
[22] J. Garriga *et al.*, JCAP **0601** (2006) 017.
[23] A. De Simone *et al.*, Phys. Rev. D **78** (2008) 063520.
[24] R. Bousso, Phys. Rev. Lett. **97** (2006) 191302.

[25] R. Bousso, B. Freivogel and I. S. Yang, Phys. Rev. D **74** (2006) 103516.

[26] G. W. Gibbons, S. W. Hawking, and J. M. Stewart, Nucl. Phys. B **281** (1987) 736.

[27] G. W. Gibbons and N. Turok, Phys. Rev. D **77** (2008) 063516.

[28] M. Li and Y. Wang, JCAP **0706** (2007) 012.

[29] M. Li and Y. Wang, JCAP **0708** (2007) 007.

[30] D. N. Page, arXiv:0707.4169.

[31] D. N. Page, Phys. Rev. D **78** (2008) 023514.

[32] J. B. Hartle and M. Srednicki, Phys. Rev. D **75** (2007) 123523.

[33] M. Li and Y. Wang, arXiv:0708.4077.

[34] R. Bousso *et al.*, Phys. Rev. D **76** (2007) 043513.

[35] L. Boltzmann, Nature **51** (1985) 413.

[36] R. Bousso *et al.*, Phys. Rev. D **83** (2011) 023525.

8
Tuning Mechanisms

As reviewed in Chapter 5, there is a no-go theorem about tuning mechanism in four spacetime dimensions. As the story of many other no-go theorems, the utility of a no-go theorem is more likely a sign to point out ways out instead of to stop. Also, as some other no-go theorems (such as the story between composed graviton and AdS/CFT), the way out is to go to extra dimensions. In this chapter we discuss some higher dimensional constructions of tuning mechanisms.

8.1 Brane versus Bulk Mechanism

We will mainly explain the work of Kachru, Schulz and Silverstein [1]. They work in brane world embedded in a 5 dimensional spacetime. They introduced a tuning scalar field, but there is always a singularity in the bulk. As Witten correctly pointed out, such a singularity can not be accepted based on general physics principle (otherwise one can introduce just about anything in our world, such as a monopole in the usual Maxwell theory).

Let us start with the action in 5 dimensions,

$$S = \int d^5x \sqrt{-G}(\mathcal{R} - \frac{4}{3}(\nabla\phi)^2 - \Lambda e^{a\phi}) + \int d^4x \sqrt{-g}(-f(\phi)), \qquad (8.1)$$

where $f(\phi) = Ve^{b\phi}$. If for any V, one can always find a flat 4 dimensional spacetime solution, then the tuning mechanism is successful. This is because V is quantum corrected (with proof of the form of $f(\phi)$ be invariant). If $f(\phi)$ is not invariant, one need to show a flat solution still exists.

The equations of motion derived from Eq. (8.1) are

$$\sqrt{-G}\left(R_{MN} - \frac{1}{2}G_{MN}\mathcal{R}\right) - \frac{4}{3}\sqrt{-G}(\nabla_M\phi\nabla_N\phi - \frac{1}{2}G_{MN}(\nabla\phi)^2)$$

$$+ \frac{1}{2}[\Lambda e^{a\phi}\sqrt{-G}G_{MN} - \sqrt{-g}g_{\mu\nu}\delta_M^\mu\delta_N^\nu\delta(x_5)] = 0\,, \tag{8.2}$$

$$\sqrt{-G}\left(\frac{8}{3}\triangle\phi - a\Lambda e^{a\phi}\right) - bV\delta(x_5)e^{b\phi}\sqrt{-g} = 0\,.$$

Let the metric be

$$ds^2 = e^{2A(x_5)}\eta_{\mu\nu}dx^\mu dx^\nu + dx_5^2\,, \tag{8.3}$$

then

$$\frac{8}{3}\phi'' + \frac{32}{3}A'\phi' - a\Lambda e^{a\phi} - bV\delta(x_5)e^{b\phi} = 0\,,$$

$$6(A')^2 - \frac{2}{3}(\phi')^2 + \frac{1}{2}\Lambda e^{a\phi} = 0\,, \tag{8.4}$$

$$3A'' + \frac{4}{3}\phi'^2 + \frac{1}{2}e^{b\phi}V\delta(x_5) = 0\,.$$

To solve the above equations, let $A' = \alpha\phi'$, consider the following cases separately.

• $\Lambda = 0$.

When $x_5 \neq 0$, from

$$6\alpha^2\phi'^2 = \frac{2}{3}\phi'^2\,, \tag{8.5}$$

we deduce $\alpha = \pm 1/3$ and

$$\phi'' \pm \frac{4}{3}\phi'^2 = 0\,. \tag{8.6}$$

Thus

$$\phi = \pm\frac{3}{4}\ln\left|\frac{4}{3}x_5 + c\right| + d\,. \tag{8.7}$$

This solution has a singularity at $x_5 = -3c/4$.

Solution 1, let

$$\alpha = \begin{cases} \dfrac{1}{3}, & x_5 < 0, \\[2mm] -\dfrac{1}{3}, & x_5 > 0, \end{cases} \tag{8.8}$$

thus

$$\phi(x_5) = \begin{cases} \phi_1(x_5) = \dfrac{3}{4} \ln \left| \dfrac{4}{3}x_5 + c_1 \right| + d_1, & x_5 < 0 \, ; \\[3mm] \phi_2(x_5) = -\dfrac{3}{4} \ln \left| \dfrac{4}{3}x_5 + c_2 \right| + d_2, & x_5 > 0 \, . \end{cases} \tag{8.9}$$

The continuity at x_5 requires

$$\frac{3}{4} \ln |c_1| + d_1 = -\frac{3}{4} \ln |c_2| + d_2 \, . \tag{8.10}$$

From equations of motion we obtain

$$\frac{2}{c_2} = \left(-\frac{3b}{8} - \frac{1}{2} \right) V e^{bd_1} |c_1|^{\frac{3}{4}b} \, ,$$
$$\frac{2}{c_1} = \left(-\frac{3b}{8} + \frac{1}{2} \right) V e^{bd_1} |c_1|^{\frac{3}{4}b} \, . \tag{8.11}$$

The solution for the parameters always exists no matter what value of V is. For an arbitrary $f(\phi)$, we have

$$\frac{8}{3}(\phi_2'(0) - \phi_1'(0)) = f'(\phi(0)) \, ,$$
$$3 \left(-\frac{1}{3}\phi_2'(0) - \frac{1}{3}\phi_1'(0) \right) = -\frac{1}{2}f(\phi(0)) \, , \tag{8.12}$$

or

$$-\frac{8}{3}(c_1^{-1} + c_2^{-1}) = f' \left(\frac{3}{4} \ln |c_1| + d_1 \right) \, ,$$
$$c_2^{-1} - c_1^{-1} = -\frac{1}{2}f \left(\frac{3}{4} \ln |c_1| + d_1 \right) \, . \tag{8.13}$$

Solution to these equations always exists.

Since the effective Newtonian constant is

$$G \sim \int dx_5 e^{2A(x_5)} \sim \int dx_5 \left| \frac{4}{3}x_5 + c_1 \right|^{\frac{1}{2}} \, . \tag{8.14}$$

This constant diverges if there is no cutoff on x_5, this is so when $c_1 < 0$ for $x_5 < 0$, thus we let $c_1 > 0$, then the lower limit of the integral of x_5 is $\underline{x_5} = -3c_1/4$. This lower limit is a singularity, however. Similarly, for $x_5 > 0$, $c_2 < 0$, there is also a singularity.

For an arbitrary function $f(\phi)$, we require

$$f(\phi(0)) > 0, \quad -\frac{4}{3} < \frac{f'}{f} < \frac{4}{3}. \tag{8.15}$$

• Fluctuations.
Let

$$g_{\mu\nu} = e^{2A}\eta_{\mu\nu} + h_{\mu\nu}, \tag{8.16}$$

then

$$h_{\mu\nu} \propto \left| \frac{4}{3}x_5 + c \right|^{\frac{1}{2}}, \tag{8.17}$$

there is also a singularity, although $h_{\mu\nu} \to 0$, but

$$x_5 \to \begin{cases} -\dfrac{3}{4}c_1, \ \phi \to -\infty \ \text{weak coupling}; \\[2mm] -\dfrac{3}{4}c_2, \ \phi \to \infty \ \text{strong coupling}. \end{cases} \tag{8.18}$$

Solution 2, in this case $\alpha_1 = \alpha_2$,

$$\phi(x_5) = \begin{cases} \pm\dfrac{3}{4}\ln\left|\dfrac{4}{3}x_5 + c_1\right| + d_1, \ x_5 < 0, \\[3mm] \pm\dfrac{3}{4}\ln\left|\dfrac{4}{3}x_5 + c_2\right| + d_2, \ x_5 > 0, \end{cases} \tag{8.19}$$

and $b = \mp\frac{4}{3}$. Or

$$f'(\phi(0)) = \mp\frac{4}{3}f(\phi(0)). \tag{8.20}$$

One finds

$$c_1 = -c_2 = c, \quad d_1 = d_2 = d, \quad e^{\mp\frac{4}{3}d} = \frac{4}{V}\frac{c}{|c|}. \tag{8.21}$$

• $\Lambda \neq 0$.
In this case $\alpha = -\frac{8}{9a}$, and

$$\phi = -\frac{2}{a}\ln\left[\frac{a \mp \sqrt{B}}{2}x_5 + d\right], \quad B = \frac{\Lambda}{\frac{4}{3} - 12\alpha^2}. \tag{8.22}$$

The junction condition leads to

$$V = -12\alpha\sqrt{B}\,, \qquad b = \frac{4}{9\alpha}\,, \tag{8.23}$$

where V is a function of a and Λ, and is fine tuned. This is similar to the Randall–Sundrum scenario.

When $a = 0$ ($\Lambda \neq 0$), let $h = \phi'$, $g = A'$, we have

$$h' + 4hg = 0\,,$$

$$6g^2 - \frac{2}{3}h^2 + \frac{1}{2}\Lambda = 0\,, \tag{8.24}$$

$$3g' + \frac{4}{3}h^2 = 0\,.$$

The case $\Lambda = 0$ may be justified in string theory provided there is supersymmetry in the 5D bulk. It is expected that we leave $f(\phi)$ to be a general function due to quantum corrections.

The major problem of this new tuning mechanism is the existence of singularities. The difference of this mechanism from that of Randall–Sundrum is that we need to fine tune relationship between the 4D cosmological constant and the 5D cosmological constant in the latter mechanism, while here the scalar field ϕ massages the corrections to V to the bulk.

There are also a number of other brane world approaches to dark energy. For example, [2, 3] considered codimension two branes. Other approaches include [4].

8.2   Black Hole Self-Adjustment

This mechanism is due to Csaki, Erlich and Grojean [5].

Introducing a black hole in the 5D bulk with metric

$$ds^2 = -h(r)dt^2 + \frac{r^2}{l^2}d\Sigma_k^2 + h^{-1}(r)dr^2\,, \tag{8.25}$$

where

$$h(r) = k + \frac{r^2}{l^2} - \frac{\mu}{r^2}\,, \tag{8.26}$$

and $d\Sigma_k^2$ is the metric on the maximally symmetric space with spatial curvature k. Now suppose there is a three-brane, we need to require a Z_2

symmetry (so that there is one black hole to the left and another black hole to the right), for $r < r_0$, the metric is given by the functions $h(r)$, r^2 and $h^{-1}(r)$, and for $r > r_0$,

$$\tilde{h}(r) = h\left(\frac{r_0^2}{r}\right), \qquad \tilde{h}^{-1}(r) = h^{-1}\left(\frac{r_0^2}{r}\right)\frac{r_0^2}{r^2}, \qquad \tilde{r} = \left(\frac{r_0^2}{r}\right)^2, \quad (8.27)$$

namely, the metric when $r > r_0$ is

$$ds^2 = -h\left(\frac{r_0^2}{r}\right)dt^2 + \left(\frac{r_0^2}{r}\right)^2 d\Sigma_1^2 + h^{-1}\left(\frac{r_0^2}{r}\right)\frac{r_0^2}{r^2}dr^2. \quad (8.28)$$

If the brane location $r_0 = R(t)$ is a function of time, the equation of motion

$$\dot{\rho} + 3(\rho + p)\frac{\dot{R}}{R} = 0 \quad (8.29)$$

has a static solution $\dot{R} = \ddot{R} = \dot{\rho} = 0$. The junction conditions are

$$6\sqrt{h(r_0)} = \kappa_5^2 \rho r_0, \qquad 18h'(r_0) = -\kappa_5^4(2 + 3w)\rho^2 r_0, \quad (8.30)$$

where $w = p/\rho$. We have

$$\mu = -\frac{1}{24}\kappa_5^4(1 + w)\rho^2 r_0^2. \quad (8.31)$$

If $\mu < 0$, there is a naked singularity in the black hole solution. So we require $\mu > 0$ thus $w < -1$ and the positive energy condition is violated. Also

$$\rho = -\frac{72}{1 + 3w}\frac{1}{l\kappa_5^2}, \quad (8.32)$$

this is fine-tuning. In order to avoid these drawbacks, we need to introduce new parameters. Assume that this is a 5D Abelian gauge field A_M and A_r is even under Z_2 and A_μ odd under Z_2. The black hole solution with charge Q is

$$h(r) = k + \frac{r^2}{l^2} - \frac{\mu}{r^2} + \frac{Q^2}{r^4},$$

$$Q^4 < \frac{4}{27}\mu^3 l^2. \quad (8.33)$$

The junction conditions are

$$36 \left(\frac{r_0^2}{l^2} - \frac{\mu}{r_0^2} + \frac{Q^2}{r_0^4} \right) = \kappa_5^4 \rho^2 r_0^2 \,,$$

$$36 \left(\frac{r_0^2}{l^2} + \frac{\mu}{r_0^2} - \frac{2Q^2}{r_0^4} \right) = -\kappa_5^4 (2 + 2w) \rho^2 r_0^2 \,. \tag{8.34}$$

The solution to these equations is

$$\mu = 3 \left(l^{-2} + \frac{1}{36} \kappa_5^4 w \rho^2 \right) r_0^4 \,,$$

$$Q^2 = 2 \left(l^{-2} + \frac{1}{72} \kappa_5^4 (1 + 3w) \rho^2 \right) r_0^6 \,. \tag{8.35}$$

The condition $Q^2 \geqslant 0$ leads to

$$\rho \leqslant \rho_0 = \sqrt{\frac{-72}{1 + 3w}} \frac{1}{l \kappa_5^2} \,, \qquad w < -\frac{1}{3} \,. \tag{8.36}$$

The condition for the existence of a horizon is $w > 0$ or $w < -1$, and $\rho > \rho_-$ where

$$\rho_- = \frac{6}{l \kappa_5^2} \left(\frac{1}{8w^3} (1 + 6w - 3w^2 + \sqrt{(1+w)^3(1+9w)}) \right)^{\frac{1}{2}} \,. \tag{8.37}$$

When $w < -1$, $\rho_- < \rho < \rho_0$, we find

$$S = \int_{r_H}^{r_0} dr \sqrt{-g} \left[\frac{1}{2\kappa_5^2} \mathcal{R} - \frac{1}{4} F^2 - \Lambda + \mathcal{L}_{\text{matter}} \delta(\sqrt{g_{rr}}(r - r_0)) \right] \,. \tag{8.38}$$

Since

$$\mathcal{L}_{\text{matter}} = p = \left(-\frac{h'}{\kappa_5^2 \sqrt{h}} - \frac{4\sqrt{h}}{\kappa_5^2 r_0} \right) \Bigg|_{r=r_0} \,,$$

$$\mathcal{R} = -h''(r) - 6\frac{h'}{h} - 6\frac{h}{r^2} + \left(2h' + 12\frac{h}{r} \right) \delta(r - r_0) \,, \tag{8.39}$$

we have

$$S = \kappa_5^{-2} r_H^2 h(r_H) = 0 \,, \tag{8.40}$$

this tells us that the 4D effective cosmological constant is zero, $\Lambda_{\text{eff}} = 0$.

• Other cases.

Take $r_0 = R(t)$ and

$$
R(t) = \begin{cases}
R_0 e^{Ht}, & k = 0, \\
\operatorname{Sinh}(Ht)/H, & k = -1, \\
\operatorname{Cosh}(Ht)/H, & k = 1,
\end{cases}
\tag{8.41}
$$

when $\Lambda_4 > 0$ and

$$
R(t) = \cos(Ht)/H,
\tag{8.42}
$$

when $\Lambda_4 < 0$. The junction conditions are

$$
\frac{\dot{R}^2}{R^2} = \frac{1}{36}\kappa_5^2 \rho^2 - \left(\frac{k}{R^2} + l^{-2} - \frac{\mu}{R^4} + \frac{Q^2}{R^6} \right),
\tag{8.43}
$$

$$
\dot{\rho} + 3H(1+w)\rho = 0.
$$

For $w = -1$, $\dot{\rho} = 0$.

(a) de Sitter.

We have $w = -1$ and

$$
\mu = 0, \qquad Q^2 = 0,
$$

$$
\frac{1}{36}\kappa_5^2 \rho^2 - l^{-2} = H^2 > 0.
\tag{8.44}
$$

(b) Anti-de Sitter.

When $w = -1$,

$$
H^2 = l^{-2} - \frac{1}{36}\kappa_5^2 \rho^2 > 0,
$$

$$
\mu = Q^2 = 0.
\tag{8.45}
$$

When $w = -\frac{1}{3}$,

$$
\mu = -\frac{1}{36}\kappa_5^4 \rho^2, \qquad Q = 0,
$$

$$
H^2 = l^{-2}.
\tag{8.46}
$$

When $w = 0$,

$$
\mu = 0, \qquad Q^2 = \frac{1}{36}\kappa_5^2 l^6 \rho_0^2.
\tag{8.47}
$$

For all the above cases a fine-tuning is required.

Finally, we note that Lorentz symmetry is violated. Since the metric is

$$ds^2 = -hdt^2 + \frac{r^2}{l^2}d\Sigma_k^2 + h^{-1}dr^2 \,, \tag{8.48}$$

the speed of gravitational wave is not equal to c, the speed of light. On the brane, the speed of light is $c = \frac{dx}{dt} = \frac{\sqrt{h}l}{r}$. When transverse to the brane, the speed of light is $\frac{dr}{dt} = h$. Detailed calculation shows that the speed of gravitational wave depends on the parameter $E/|\vec{p}|$, where E and $\vec{p}$ are conserved quantities.

References

[1] S. Kachru, M. B. Schulz, and E. Silverstein, Phys. Rev. D **62** (2000) 045021.

[2] J. M. Cline *et al.*, JHEP **0306** (2003) 048.

[3] J. Vinet and J. M. Cline, Phys. Rev. D **70** (2004) 083514.

[4] I. P. Neupane, Phys. Rev. D **83** (2011) 086004; I. P. Neupane, Int. J. Mod. Phys. D **19** (2010) 2281.

[5] C. Csaki, J. Erlich, and C. Grojean, Nucl. Phys. B **604** (2001) 312.

9
Modified Gravity

Gravity, especially in its quantum aspects, encounters a number of puzzles. Facing those puzzles, it is a long standing question whether to understand or to modify gravity. Modified gravity is now a huge category. Here we shall choose to introduce several classes of models most related to dark energy, namely $f(\mathcal{R})$, MOND, DGP type models and briefly mention other directions.

9.1 $f(\mathcal{R})$ Models

Well before dark energy or inflation has been proposed, there have been already attempts to replace the Ricci scalar $\mathcal{R}$ in the gravitational action by a general function $f(\mathcal{R})$ [1].

It is well known that for the standard Einstein–Hilbert action, one can either treat metric itself as dynamical variables, or treat both metric and connection as variables when doing variation. The former is known as the metric formulation, and the latter is known as the Palatini formulation [2, 3].[1]

To generalize to $f(\mathcal{R})$, there are thus two possibilities, the metric generalization or the Palatini generalization. Turns out that these two are not identical for $f(\mathcal{R})$. Here we shall mainly discuss the metric formulation. The readers interested in the Palatini formulation are referred to the review [5].

It is also helpful to note that these $f(\mathcal{R})$ models can be related to standard gravity with a non-minimally coupled scalar field, using conformal

---

[1] The so-called Palatini formulation of general relativity was perhaps also discovered by Einstein [3]. See a historical review [4] for the story for readers who do not speak German or Italian.

transformation [6, 7]. A conformal transformation is not a symmetry of general relativity nor its $f(\mathcal{R})$ generalization. However there are ways to match observables such that calculation on one side can be used on the other side. For these aspects, the readers are referred to the review [8].

Now we shall review the $f(\mathcal{R})$ gravity models applied to dark energy.[2] To be specific, we here review two simple models:

• The CDTT model.

The whole thing started with the paper of Carroll, Duvvuri, Trodden and Turner [10]. The idea is to use other Lagrangian terms to generate accelerated solutions. Although these terms are more complicated and more unnatural than the Einstein's cosmological term, we can not logically rule out these possibilities and need to test them. The simplest example is

$$S = \frac{1}{16\pi G} \int d^4x \sqrt{-g} \left(\mathcal{R} - \frac{\mu^4}{\mathcal{R}} \right) + S_{\text{matter}} \,. \tag{9.1}$$

The equation of motion is

$$\left(1 + \frac{\mu^4}{\mathcal{R}^2} \right) R_{\mu\nu} - \frac{1}{2} \left(1 - \frac{\mu^4}{\mathcal{R}^2} \right) g_{\mu\nu} \mathcal{R}$$

$$+ \mu^4 [g_{\mu\nu} - \nabla_{(\mu} \nabla_{\nu)}] \mathcal{R}^{-2} = 8\pi G T_{\mu\nu} \,. \tag{9.2}$$

Consider matter as an ideal fluid $T_{\mu\nu} = (\rho + p)u_\mu u_\nu + p g_{\mu\nu}$. Since

$$\mathcal{R} = 6 \left[\frac{\ddot{a}}{a} + \left(\frac{\dot{a}}{a} \right)^2 \right] = 6 \left(\dot{H} + 2H^2 \right) , \tag{9.3}$$

we obtain the modified Friedmann equation

$$3M_{\text{P}}^2 H^2 - \frac{\mu^4 M_{\text{P}}^2}{12(\dot{H} + 2H^2)^3} (2H\ddot{H} + 15H^2\dot{H} + 2\dot{H}^2 + 6H^4) = \rho \,. \tag{9.4}$$

The other Friedmann equation is not independent and can be obtained from the above equation and the continuity of equation of ρ.

We make a field redefinition

$$e^{\alpha\varphi} = 1 + \frac{\mu^4}{\mathcal{R}^2} \,,$$

$$\tilde{g}_{\mu\nu} = e^{\sqrt{\frac{2}{3}} M_{\text{P}}^{-1} \varphi} g_{\mu\nu} = e^{\alpha\varphi} g_{\mu\nu} \,, \tag{9.5}$$

---

[2] Here we just introduce the four-dimensional $f(\mathcal{R})$ models. For five-dimensional $f(\mathcal{R})$ gravity model, see [9].

and

$$\tilde{T}_{\mu\nu} = (\tilde{\rho} + \tilde{p})\tilde{u}_\mu \tilde{u}_\nu + \tilde{p}\tilde{g}_{\mu\nu}, \tag{9.6}$$

where

$$\tilde{\rho} = e^{-2\alpha\varphi}\rho, \qquad \tilde{p} = e^{-2\alpha\varphi}p, \qquad \tilde{u}_\mu = e^{\frac{1}{2}\alpha\varphi}u_\mu, \tag{9.7}$$

then

$$S(\tilde{g}, \varphi) = \frac{1}{16\pi G} \int d^4x \sqrt{-\tilde{g}}\tilde{R} + \int d^4x \sqrt{-\tilde{g}} \left[-\frac{1}{2}(\partial\varphi)^2 - V(\varphi) \right]$$
$$+ S_{\text{matter}}(\tilde{g}, \varphi), \tag{9.8}$$

where

$$V(\varphi) = \mu^2 M_{\text{P}}^2 e^{-2\alpha\varphi} \sqrt{e^{\alpha\varphi} - 1}. \tag{9.9}$$

Thus, the modified theory is equivalent to a tensor-scalar theory, although matter is not canonically coupled to the new metric. The new cosmological equations are

$$3\tilde{H}^2 = M_{\text{P}}^{-2}(\tilde{\rho} + \rho_\varphi),$$
$$\ddot{\varphi} + 3\tilde{H}\dot{\varphi} + V' - 1 - 3w\sqrt{6}\tilde{\rho} = 0. \tag{9.10}$$

The last term in the equation of motion of φ arises from the coupling between matter and φ. Thus, the theory is equivalent to the theory of quintessence with coupling to matter.

The evolution of matter is

$$\tilde{\rho} = \tilde{\rho}_0 \tilde{a}^{-3(1+w)} e^{\frac{3w-1}{\sqrt{6}} M_{\text{P}}^{-1} \varphi}. \tag{9.11}$$

There are two separated cases.

(a) Eternal de Sitter, $V' = 0$, but this solution is unstable.

(b) Power-law acceleration.

When $e^{\alpha\varphi} \gg 1$, $V \sim \mu^2 M_{\text{P}}^2 e^{-\sqrt{\frac{3}{2}} M_{\text{P}}^{-1} \varphi}$, then

$$\tilde{a}(\tilde{t}) \propto \tilde{t}^{\frac{4}{3}}, \qquad a(t) \propto t^2. \tag{9.12}$$

The problem of this model is obvious, it is similar to the Brans–Dicke theory thus in general violates the equivalence principle.

Chiba [11] considered more generally the model

$$S = \frac{1}{2}M_{\rm P}^2 \int d^4x \sqrt{-g} f(\mathcal{R}) + S_{\rm matter}(g) \,, \qquad (9.13)$$

which is equivalent to

$$S = \frac{1}{2}M_{\rm P}^2 \int d^4x \sqrt{-g} \left[f(\varphi) + f'(\varphi)(\mathcal{R} - \varphi) \right] . \qquad (9.14)$$

The variation of the above action with respect to φ leads to $\varphi = \mathcal{R}$ if $f'(\varphi) \neq 0$.

Redefine

$$\tilde{g}_{\mu\nu} = f'(\varphi) g_{\mu\nu} \,, \qquad (9.15)$$

then

$$S = \frac{1}{2}M_{\rm P}^2 \int d^4x \sqrt{-\tilde{g}} \left[\tilde{R} - \frac{3}{2f'^2}(\partial\varphi)^2 - \frac{1}{f'^2}(\varphi f' - f) \right] + S_{\rm matter} \,. \quad (9.16)$$

As for the Brans–Dicke theory, we require

$$\gamma - 1 < 2.8 \times 10^{-4} \,, \qquad \gamma = \frac{\omega + 1}{\omega + 2} \,. \qquad (9.17)$$

For the CDTT model, $\omega = 0$, so the equivalence principle is violated. For the Starobinsky model $f(\mathcal{R}) = \mathcal{R} + M^{-2}R^2$, we need $M \sim 10^{12}$ GeV, but the constraint $|\gamma - 1| < 2.8 \times 10^{-4}$ is derived for $m \leqslant 10^{-27}$ GeV.

• Modified source gravity.

To avoid the above problem, a new model was proposed [12]. Again, the idea of conformal transformation is used. Let $\psi = \frac{1}{2}\ln f'(\varphi)$, and

$$\tilde{g}_{\mu\nu} = e^{2\psi} g_{\mu\nu} \,, \qquad (9.18)$$

then the action becomes

$$\int d^4x \sqrt{-\tilde{g}} \left[\frac{1}{2}M_{\rm P}^2 \tilde{\mathcal{R}} - 3\tilde{g}^{\mu\nu}(\nabla\psi)^2 - V(\psi) \right]$$

$$+ \int d^4x \mathcal{L}(e^{-2\psi}\tilde{g}_{\mu\nu}, \chi_{\rm m}) \,, \qquad (9.19)$$

where $\chi_{\rm m}$ denotes matter fields collectively, and

$$V(\psi) = \frac{\varphi f'(\varphi) - f(\varphi)}{2f(\varphi)} M_{\rm P}^2 \,. \qquad (9.20)$$

There is still a problem to be consistent with the solar system. To avoid this problem, some introduced Palatini formalism, or remove the kinetic term of ψ. The model without the kinetic term of ψ is the modified source gravity theory. ψ in this model becomes a Lagrangian multiplier,

$$S = \int d^4x\sqrt{-\tilde{g}}\left[\frac{1}{2}M_{\rm P}^2\tilde{\mathcal{R}} - V(\psi)\right] + S_{\rm m}(e^{-2\psi}\tilde{g},\chi_{\rm m})\,. \tag{9.21}$$

When there is no matter, ψ is just number, and $V(\psi)$ becomes a constant. Rescaling back to $g_{\mu\nu}$,

$$S = \int d^4x\sqrt{-g}\left[\frac{1}{2}M_{\rm P}^2e^{2\psi}\mathcal{R} + 3e^{2\psi}(\partial\psi)^2 - e^{4\psi}V(\psi)\right]$$
$$+ S_{\rm m}(g,\chi_{\rm m})\,. \tag{9.22}$$

In the above action, the coupling between $g_{\mu\nu}$ and matter is simpler, and there is a superficial kinetic term for ψ. The Einstein equation is

$$M_{\rm P}^2e^{2\psi}G_{\mu\nu} = T_{\mu\nu}(\chi) + T_{\mu\nu}(\psi)\,, \tag{9.23}$$

where

$$T_{\mu\nu}(\psi) = -2\partial_\mu\psi\partial_\nu\psi + 2\nabla_\mu\nabla_\nu\psi$$
$$- g_{\mu\nu}[e^{4\psi}V(\psi) + (\partial\psi)^2 + 2\nabla\triangle\psi]\,. \tag{9.24}$$

The equation of motion for ψ is

$$\triangle\psi + (\partial\psi)^2 + \frac{1}{6M_{\rm P}^2}e^{-2\psi}\partial_\psi(e^{4\psi}V) - \frac{1}{6}\mathcal{R} = 0\,. \tag{9.25}$$

In the following, we use $U(\psi) = e^{4\psi}V(\psi)$. There appears kinetics for ψ, but, taking the trace of the Einstein equation we have

$$\frac{1}{6}\mathcal{R} = \frac{e^{-2\psi}}{6M_{\rm P}^2}(-T + 4U) + (\partial\psi)^2 + \triangle\psi\,, \tag{9.26}$$

this combined with the EOM for ψ we find

$$U'(\psi) - 4U(\psi) = -T\,, \tag{9.27}$$

this is an algebraic equation. We consider several cases in the following.[3]

---

[3] The thermal dynamical properties of $f(\mathcal{R})$ gravity is different from Einstein gravity, which was discussed in [13].

(a) The solar system.

In this case $T_{\mu\nu}(\chi_{\mathrm{m}}) = 0$, thus

$$4U(\psi_0) - U'(\psi_0) = 0 \,, \tag{9.28}$$

the contribution of ψ_0 to $T_{\mu\nu}$ vanishes, so

$$M_{\mathrm{P}}^2 G_{\mu\nu} = -e^{2\psi_0} U(\psi_0) g_{\mu\nu} \,. \tag{9.29}$$

This is the Einstein equation with a cosmological constant, provided the cosmological constant is small enough, the solar system is fine.

(b) Interaction with matter.

Integrating out ψ results in a complicated and non-renormalizable action, so it is better to view the model as a low energy effective model.

(c) Cosmology.

Let $T_{\nu}^{\mu} = \mathrm{diag}(-\rho, p, p, p)$, the Friedmann equation is

$$3H^2 + \frac{3k}{a^2} = M_{\mathrm{P}}^{-2} e^{-2\psi} [\rho + U(\psi)] - 3\dot{\psi}^2 - 6H\dot{\psi} \,. \tag{9.30}$$

Let $x = \ln a$, $\dot{\psi} = H\psi_{,x}$,

$$H^2 = (1 + \psi_{,x})^{-2} \left[\frac{1}{3M_{\mathrm{P}}^2} e^{-2\psi} (\rho + U(\psi)) - \frac{k}{a^2} \right] \,. \tag{9.31}$$

We see that the effective Newtonian constant is $8\pi G_{\mathrm{eff}} = M_{\mathrm{P}}^{-2} e^{-2\psi} (1 + \psi_{,x})^{-2}$. We have the freedom to choose the form of $U(\psi)$, amounting to choosing $\psi(\rho)$. Let

$$e^{-4\psi} = \alpha \frac{\rho_0}{\rho} + 1 \,, \tag{9.32}$$

so when $\rho \to \infty$, $\psi \to 0$ and when $\rho \to 0$, $\psi \to -\infty$. Since $U' - 4U = -T = \rho$, thus

$$U(\rho) = -\frac{\alpha\rho_0\rho}{4(\alpha\rho_0 + \rho)} \ln\left(\frac{a\rho_0}{\rho}\right) \,,$$

$$U(\psi) = \alpha\rho_0 e^{4\psi} \left[\psi - \frac{1}{4}\ln(1 - e^{4\psi}) \right] \,, \quad \psi_{,x} = -\frac{3\alpha\rho_0}{4(\alpha\rho_0 + \rho)} \,. \tag{9.33}$$

The Friedmann equation is

$$H^2 = \left(\frac{4\alpha\rho_0 + 4\rho}{\alpha\rho_0 + 4\rho}\right)^2 \frac{\sqrt{\rho}}{3M_P^2} \frac{\alpha\rho_0 + \rho - \frac{\alpha\rho_0}{4}\ln\frac{\alpha\rho_0}{\rho}}{\sqrt{\alpha\rho_0 + \rho}} - \frac{k}{a^2}. \tag{9.34}$$

Let $\rho_{\rm de} = 3M_P^2 H^2 - \rho$, then

$$w_{\rm de} = -1 - \frac{1}{3}\frac{d\ln\rho_{\rm de}}{d\ln a}. \tag{9.35}$$

Using the above equations in fitting the data, one finds

$$\alpha = 0.98, \qquad h = 0.72. \tag{9.36}$$

9.2   MOND and TeVeS Theories

The MOND theory (the modified Newtonian dynamics) was proposed by Milgrom as a substitute for dark matter [14]. It proposes to modify Newtonian dynamics at very large scales in order to explain the rotational curves of galaxies.

Since MOND is aimed to modify Newtonian dynamics, the most appropriate starting point is an equation for the Newtonian potential. The typical acceleration involved in the rotational curves is of order $a_0 = 10^{-8}$ cm/s^2. To make use of this fact, the basic assumption of MOND is the equation

$$\mu(a/a_0)a = -\nabla\Phi, \tag{9.37}$$

where Φ is the Newtonian potential, $\mu(a/a_0)$ is a function with the property $\mu \to 1$ when $a \gg a_0$. If $\mu(x) = x$ for $x \ll 1$, then we have

$$\frac{a^2}{a_0} = -\nabla\Phi = \frac{GM}{r^2}. \tag{9.38}$$

Since $a = v^2/r$, the above leads to $v = $ const.

Bekenstein and Milgrom [15] found that the MOND theory coupled to matter can be derived from an AQUadratic Lagrangian (AQUAL)

$$\mathcal{L} = -\frac{a_0^2}{8\pi G}f\left(\frac{(\nabla\Phi)^2}{a_0^2}\right) - \rho\Phi, \tag{9.39}$$

where f is given by $f'(x) = \mu(\sqrt{x})$. The equation of motion is

$$\nabla \cdot \left(\mu\left(\frac{\nabla\Phi}{a_0}\right)\nabla\Phi\right) = 4\pi G\rho, \tag{9.40}$$

with

$$
f(y) = \begin{cases} y, & y \gg 1; \\ \dfrac{2}{3} y^{\frac{3}{2}}, & y \ll 1. \end{cases}
$$
(9.41)

• Relativistic MOND.

The MOND action AQUAL can be easily generalized into the relativistic case [15]. Again a scalar field Ψ is introduced in this model. Matter is coupled to $e^{2\Psi} g_{\mu\nu}$, the Lagrangian of Ψ and the action of a particle in the background are

$$
\mathcal{L}_\Psi = -\frac{1}{8\pi GL^2} \tilde{f}(L^2 g^{\mu\nu} \partial_\mu \Psi \partial_\nu \Psi)\,,
$$
$$
S_{\mathrm{m}} = -m \int e^\Psi \sqrt{-g_{\mu\nu} \dot{x}^\mu \dot{x}^\nu}\, dt\,.
$$
(9.42)

The relativistic MOND model is also sometimes called relativistic AQUAL, or simply AQUAL. In the low velocity limit, the action of a particle is

$$
e^\Psi ds \sim \left(1 + \Phi - \Psi - \frac{v^2}{2}\right) dt\,,
$$
(9.43)

thus

$$
a = -\nabla(\Phi + \Psi)\,.
$$
(9.44)

Combined with the action of Ψ, we obtain MOND.

There are two problems with relativistic MOND:

(a) When $\tilde{f} = f$, causality is violated. The discussion of causality and related modification of MOND, can be found in [16].

(b) Light is almost decoupled from Ψ, dark matter deflects light more than relativistic MOND does. This unfortunately conflicts with the gravitational lensing experiments. Thus the relativistic MOND is ruled out.

• PCG (The phase coupled gravity).

In addition to Ψ, another field A is introduced with Lagrangian [17, 18]

$$
\mathcal{L}(\Psi, A) = -\frac{1}{2}[A_{,\mu} A^{,\mu} + \eta^{-2} A^2 \Psi_{,\mu} \Psi^{,\mu}) + V(A^2)]\,,
$$
(9.45)

the equation of motion is

$$
\triangle A - \eta^{-2} A (\partial \Psi)^2 - A V'(A) = 0\,,
$$
$$
\nabla_\mu(A^2 \nabla^\mu \Psi) = \eta^2 e^\Psi M \delta(x)\,.
$$
(9.46)

If η is very small, this model reduces to relativistic MOND, but without violation of causality. Still, it contradicts gravitational lensing experiments.

• Deformed metric.

Since both relativistic MOND and PCG contradict gravitational lensing experiments, a deformed metric theory was introduced [19]. Let

$$\tilde{g}_{\mu\nu} = e^{-2\Psi}(A(X)g_{\mu\nu} + B(X)\partial_\mu\Psi\partial_\nu\Psi),$$

$$X = -\frac{1}{2}(\partial\Psi)^2. \tag{9.47}$$

The second is introduced to deflect light more. To make sure of causality is not violated, the sign of B is chosen such that the deflection of light is not enough.

It is thus motivated to introduce U_μ to replace $\Psi_{,\mu}$ and

$$\tilde{g}_{\mu\nu} = e^{-2\Psi}g_{\mu\nu} - 2U_\mu U_\nu \sinh(2\Psi), \tag{9.48}$$

where U_μ is timelike.

• TeVeS.

TeVeS is by far the most complicated theory among the MOND like models [20]. Besides the metric, there are new fields φ, σ, U_μ, with $U^2 = -1$ and

$$\tilde{g}_{\mu\nu} = e^{-2\varphi}g_{\mu\nu} - 2U_\mu U_\nu \sinh 2\varphi, \tag{9.49}$$

and actions

$$S = \frac{1}{16\pi G}\int d^4x\sqrt{-g}\mathcal{R},$$

$$S(\varphi,\sigma) = -\frac{1}{2}\int d^4x[\sigma^2 h^{\mu\nu}\varphi_{,\mu}\varphi_{,\nu} + \frac{1}{2}Gl^{-2}F(k\sigma)]\sqrt{-g},$$

$$h^{\mu\nu} = g^{\mu\nu} - U^\mu U^\nu, \tag{9.50}$$

$$S(U) = -\frac{K}{32\pi G}\int d^4x[U_{[\mu,\nu]}U^{[\mu,\nu]} - \frac{2\lambda}{K}(U^2+1)]\sqrt{-g}.$$

The dynamics of $g_{\mu\nu}$ remains unchanged, however matter is minimally coupled to $\tilde{g}_{\mu\nu}$, for example

$$-m\int d^4x\sqrt{-\tilde{g}_{\mu\nu}\dot{x}^\mu\dot{x}^\nu}\,dt. \tag{9.51}$$

The equations of motion

$$G_{\mu\nu} = 8\pi G(\tilde{T}_{\mu\nu} + (1 - e^{-4\varphi}U^\alpha \tilde{T}_{\alpha(\mu} U_{\nu)} + \tau_{\mu\nu}) + \Theta_{\mu\nu} ,$$

$$\tau_{|m\mu\nu} = \sigma^2(\varphi_{,\mu}\, \varphi_{,\nu} + \cdots) , \tag{9.52}$$

$$\Theta_{\mu\nu} = K[g^{\alpha\beta}U_{[\mu,\alpha]}U_{[\nu,\beta]} + \cdots] .$$

In TeVeS, σ is a Lagrangian multiplier. This model can avoid problems of violating causality and contradicting gravitational lensing experiments, and reduces to MOND in a certain limit. Taking a proper function F, it can also generate the effect of dark energy.

The major problem of TeVeS is that if $\tilde{g}_{\mu\nu}$ is taken as the physical metric, then the tensor, vector and scalar perturbations always have propagation speed greater than speed of light. Also, the appearance of dark energy in this model is unnatural.

9.3   DGP Model

This theory is actually quite simple. It postulates that there are two independent gravity theories, one on a 3+1 dimensional brane and another in the 4+1 dimensional bulk [21].[4] The action is

$$S = \frac{1}{2\kappa_5^2} \int d^4x dy \sqrt{-G} \mathcal{R} + \frac{1}{2\kappa_4^2} \int d^4x \sqrt{-g} \mathcal{R} + S_m , \tag{9.53}$$

where the metric on the brane is $g_{\mu\nu} = G_{\mu\nu}|_{\text{brane}}$. Dvali, Gabadadze and Porrati found out the 3+1 propagator. To do so, they first studied the 3+1 propagator of a scalar with action

$$S = M^3 \int d^4x dy \partial_A \phi \partial^A \phi + M_P^2 \int d^4x dy \delta(y) \partial_\mu \phi \partial^\mu \phi . \tag{9.54}$$

The Green's function satisfies

$$(M^3 \partial_A \partial^A + M_P^2 \delta(y) \partial_\mu \partial^\mu) G(x, y; 0) = \delta^4(x) \delta(y) . \tag{9.55}$$

Take $G(x, y; 0)$ as the retarded Green's function, the potential generated by the scalar is

$$V(r) = \int G_R(t, \vec{x}, y = 0; 0) dt . \tag{9.56}$$

---

[4] The idea of brane bulk energy exchange as an origin of dark energy was also considered in [22].

Let

$$G_R(x, y; 0) = \int \frac{d^4p}{(2\pi)^4} e^{ipx} \tilde{G}_R(p, y), \qquad (9.57)$$

$\tilde{G}$ satisfies

$$(M^3(p^2 - \partial_y^2) + M_P^2 p^2 \delta(y)) \tilde{G}_R(p, y) = \delta(y). \qquad (9.58)$$

In the Euclidean space (after Wick rotation), the retarded function is

$$\tilde{G}_R(p, y) = \frac{1}{M_P^2 p^2 + 2M^3 p} e^{-p|y|}, \qquad (9.59)$$

we have

$$V(r) = -\frac{1}{8\pi^2 M_P^2} \frac{1}{r} \left\{ \sin \frac{r_0}{r} \text{Ci} \left(\frac{r_0}{r} \right) + \frac{1}{2} \cos \frac{r_0}{r} \left[\pi - 2\text{Si} \left(\frac{r_0}{r} \right) \right] \right\}, \qquad (9.60)$$

where

$$\text{Ci}(z) = \gamma + \ln z + \int_0^z \frac{dt}{t} (\cos t - 1), \quad \text{Si}(z) = \int_0^z \frac{dt}{t} \sin t, \qquad (9.61)$$

where $\gamma = 0.577 \cdots$ is the Euler constant and $r_0 = M_P^2/(2M^3)$. When $r \ll r_0$,

$$V(r) = -\frac{1}{8\pi^2 M_P^2} \frac{1}{r} \left[\frac{\pi}{2} + \left(-1 + \gamma + \ln \frac{r_0}{r} \right) \frac{r}{r_0} \right] + O(r), \qquad (9.62)$$

it has the correct $1/r$ form in 3+1 dimensions. When $r \gg r_0$,

$$V(r) = -\frac{1}{8\pi^2 M_P^2} \frac{1}{r} \left[\frac{r_0}{r} + O(1/r^2) \right], \qquad (9.63)$$

it goes like $1/r^2$, the correct potential form in 4+1 dimensions.

Thus, r_0 is a crucial scale, the world behaves as a 3+1 dimensional one when r is much smaller than this scale and as a 4+1 dimensional one when r is much larger than this scale. Let $r_0 \sim 10^{28}$ cm. we have $M \sim 10^{12}$ cm$^{-1} \sim 10$ MeV.

Back to gravity, there is one problem, namely there is one more degree of freedom. Let

$$G_{AB} = \eta_{AB} + h_{AB}, \qquad (9.64)$$

and take the gauge

$$\partial^A h_{AB} = \frac{1}{2}\partial_B h\,,\tag{9.65}$$

the equation for the Green's function is

$$(M^3\partial_A\partial^A + M_P^2\delta(y)\partial_\mu\partial^\mu)h_{\mu\nu}$$

$$= M_P^2\delta(y)\partial_\mu\partial_\nu h_{55} + \left[T_{\mu\nu} - \frac{1}{2}\eta_{\mu\nu}T\right]\delta(y)\,.\tag{9.66}$$

The Fourier transform of h is

$$\tilde{h}_{\mu\nu}(p, y = 0)\tilde{T}^{\mu\nu}(p) = \frac{\tilde{T}^{\mu\nu}\tilde{T}_{\mu\nu} - \frac{1}{3}\tilde{T}\tilde{T}}{M_P^2 p^2 + 2M^3 p}\,.\tag{9.67}$$

One reads the tensor structure from the above formula,

$$D^{\mu\nu;\alpha\beta} = \frac{1}{2}\eta^{\mu\alpha}\eta^{\nu\beta} + \frac{1}{2}\eta^{\mu\beta}\eta^{\nu\alpha} - \frac{1}{3}\eta^{\mu\nu}\eta^{\alpha\beta}\,.\tag{9.68}$$

Although the tensor structure is not what we want, when $r \ll r_0$, one still gets the correct Newtonian potential and when $r \gg r_0$ one gets the Newtonian potential in 4+1 dimensions.

• Cosmology.

To study cosmology, let us consider the following form of time-dependent metric

$$ds^2 = -n^2(t,y)dt^2 + b^2(t,y)dy^2 + a^2(t,y)d\Sigma^2\,,\tag{9.69}$$

with the Einstein equation

$$G_{AB} = \kappa^2 T_{AB}\,,$$

$$T_{AB} = \mathrm{diag}(-\rho_B, p_B, \cdots, p_B) + \frac{\delta(y)}{b}\mathrm{diag}(-\rho, p, p, p, 0)\,.\tag{9.70}$$

Solving the Einstein equation with the junction conditions

$$\frac{\kappa_5^2}{2\kappa_4^2}\left(H^2 + \frac{k}{a^2}\right) - \frac{\kappa^2}{6}\rho = \epsilon\left(H^2 - \frac{\kappa^2}{2}\rho_B - \frac{C}{a^4} + \frac{k}{a^2}\right)^{\frac{1}{2}}\,,\tag{9.71}$$

where $\epsilon = \pm 1$, taking $\rho_B = C = 0$, we find

$$H^2 + \frac{k}{a^2} = \left[\frac{\epsilon}{2r_0} + \left(\frac{1}{3M_P^2}\rho + \frac{1}{4r_0^2}\right)^{\frac{1}{2}}\right]^2\,.\tag{9.72}$$

Consider two limits of the above equation:

(a) $\frac{1}{3M_P^2}\rho \gg \frac{1}{4r_0^2}$.

This is the case in the very early universe when the energy density is large. we have

$$H^2 + \frac{k}{a^2} = \frac{\rho}{3M_P^2} + \cdots . \tag{9.73}$$

This is the usual Friedmann equation.

(b) $\frac{1}{3M_P^2}\rho \ll \frac{1}{4r_0^2}$.

We have

$$H^2 + \frac{k}{a^2} = \frac{1}{4r_0^2}(1+\epsilon)^2 . \tag{9.74}$$

Take $\epsilon = 1$, then

$$H^2 + \frac{k}{a^2} = \frac{1}{r_0^2} = \frac{\rho_{de}}{3M_P^2} , \tag{9.75}$$

this is the Friedmann equation with a cosmological constant $\rho_{de} = 3M_p^2/r_0^2$, or $\Lambda^{-1} = r_0^2$. Thus, the universe is automatically accelerated in the late stage in the DGP model.

What is the physical meaning of ϵ? Consider n, a and b as functions of y near $y = 0$,

$$\begin{aligned}
n(t,y) &= 1 + \epsilon|y|\ddot{a}(\dot{a}^2 + k)^{-\frac{1}{2}} , \\
a(t,y) &= a + \epsilon|y|(\dot{a}^2 + k)^{\frac{1}{2}} , \\
b(t,y) &= 1 .
\end{aligned} \tag{9.76}$$

Let $\Omega_{r_0} = 1/(4r_0^2 H_0^2)$, $\Omega_{k0} = -k/(H_0^2 a_0^2)$, we have

$$\Omega_{k0} + (\sqrt{\Omega_{r_0}} + \sqrt{\Omega_{r_0} + \Omega_{m0}})^2 = 1 , \tag{9.77}$$

this is different from the usual Friedmann equation

$$\Omega_{k0} + \Omega_{m0} + \Omega_{de0} = 1 . \tag{9.78}$$

Ω_{r_0} can be regarded as the fraction of dark energy.

Later, Nicolis, Rattazzi and Trincherini proposed a Galileon model of gravity as a modification of DGP, aiming at curing the ghost instabilities of the DGP self-accelerating solution [23]. A simplified version of [23], as a scalar field model will be discussed in the chapter of phenomenological models.

9.4 Other Modified Gravity Theories

There are perhaps too many modified gravity theories to review here. In this section we briefly mention some other modified gravity models.

• Brans–Dicke and scalar tensor theories.

In the early 1960s, there had already been modified gravity theory known as Brans–Dicke gravity [24]. The gravitational part of the action is

$$S = \int d^4x \sqrt{-g} \left[\frac{1}{2}\varphi\mathcal{R} - \frac{\omega}{2\varphi}\partial_\mu\varphi\partial^\mu\varphi \right],$$ (9.79)

where ω is a constant. The theory is later generalized to $f(\varphi, \mathcal{R})$, known as the scalar tensor theory [25]. In addition, this theory has also been extended to the case of 5-dimensional [26].

• Gauss–Bonnet gravity and Lovelock gravity.

Gauss–Bonnet gravity [27] was proposed in the mid 1980s, aiming to derive a low energy effective gravitational action from string theory. In D dimensions, the action of Gauss–Bonnet gravity has an Einstein–Hilbert part, plus a correction

$$\int d^D x \sqrt{-g}\, \mathcal{G}, \qquad \mathcal{G} \equiv R^{\mu\nu\rho\sigma}R_{\mu\nu\rho\sigma} - 4R^{\mu\nu}R_{\mu\nu} + \mathcal{R}^2.$$ (9.80)

This is the only ghost free combination at the $\mathcal{R}^2$ order. However, in four dimensions the Gauss–Bonnet term is a total derivative. In order to have cosmological implications, in [28], Nojiri, Odintsov and Sasaki coupled the Gauss–Bonnet term $\mathcal{G}$ to a scalar field, and obtained a model of dark energy.

Well before Gauss–Bonnet gravity became known in string theory, its more general form had already been there in the early 1970s. This is known as the Lovelock gravity [29]. The Lagrangian of N-th order Lovelock gravity is the summation

$$\mathcal{L} = \sum_{m=0}^{N} c_m \mathcal{L}_m,$$ (9.81)

where $\mathcal{L}_m$ is called the Euler density:

$$\mathcal{L}_m = 2^{-m} \delta^{a_1 b_1 \cdots a_m b_m}_{c_1 d_1 \cdots c_m d_m} R^{c_1 d_1}_{a_1 b_1} \cdots R^{c_m d_m}_{a_m b_m}.$$ (9.82)

The term $m = 0$ corresponds to a cosmological constant; $m = 1$ corresponds to the Einstein–Hilbert action and $m = 2$ corresponds to the Gauss–Bonnet term.

Unfortunately, in four dimensions the $m \geqslant 3$ terms are simply zero. Thus Lovelock gravity cannot be directly applied to dark energy. Nevertheless, 3rd order Lovelock dark energy models have been proposed from a dimensional reduction [30].

• Horava–Lifshitz gravity.

One of the biggest problems in quantum gravity is that the Einstein gravity is nonrenormalizable. Horava [31] suggested that (at least power counting) renormalizability can be obtained by giving up Lorentz invariance at high energies. The idea is to introduce higher order spatial derivatives while keep the time derivative at second order in the equation of motion. For cosmological applications, Saridakis proposed a model of Horava–Lifshitz dark energy [32].

• $f(\mathcal{T})$ gravity.

Alternatively, torsion might also be useful in modified gravity. Bengochea, Ferraro and Linder [33, 34] considered the cosmology of a so-called $f(\mathcal{T})$ gravity theory. The gravitational action is

$$S = \frac{M_{\mathrm{P}}^2}{2} \int d^4x \; e\mathcal{T} \,, \tag{9.83}$$

where $e \equiv \det e^\mu{}_a$, and $e^\mu{}_a$ is the tetrad field satisfying $g_{\mu\nu}e^\mu{}_a e^\nu{}_b = \eta_{ab}$. The scalar $\mathcal{T}$ is defined as

$$\mathcal{T} \equiv S_\rho{}^{\mu\nu}\mathcal{T}^\rho{}_{\mu\nu} \,, \qquad \mathcal{T}^\rho{}_{\mu\nu} \equiv -e^\rho{}_a(\partial_\mu e^a{}_\nu - \partial_\nu e^a{}_\mu) \,, \tag{9.84}$$

$$S_\rho{}^{\mu\nu} \equiv \frac{1}{2}(K^{\mu\nu}{}_\rho + \delta^\mu{}_\rho \mathcal{T}^{\sigma\nu}{}_\sigma - \delta^\nu{}_\rho \mathcal{T}^{\sigma\mu}{}_\sigma) \,,$$

$$K^{\mu\nu}{}_\rho \equiv -\frac{1}{2}(\mathcal{T}^{\mu\nu}{}_\rho - \mathcal{T}^{\nu\mu}{}_\rho - \mathcal{T}_\rho{}^{\mu\nu}) \,. \tag{9.85}$$

It is shown that the $f(\mathcal{T})$ theory is also a candidate of dark energy [33, 34].[5]

• Conformal gravity.

Mannheim [36] suggested that the gravitational action could be modified using the conformal Weyl tensor as

$$-\alpha_g \int d^4x \sqrt{-g} C_{\mu\nu\rho\lambda} C^{\mu\nu\rho\lambda} \,. \tag{9.86}$$

---

[5] In a recent work [35], a $f(\mathcal{R}, \mathcal{T})$ gravity model, where the gravitational Lagrangian is given by an arbitrary function of $\mathcal{R}$ and of $\mathcal{T}$, is proposed.

It is shown that this model on the one hand fits well the galactic rotation curves [37] on galactic scales, on the other hand can behave as a component of dark energy on cosmological scales [38].

• Fat graviton.

Sundrum [39] proposed a scenario of fat graviton, in which the graviton has a size instead of localized. Unlike the other approaches where an action is proposed at the first place, the fundamental principle of a fat graviton scenario is yet unknown. Instead, the work [39] aimed to address how can a graviton be fat, without violating known physics. Especially, with the help of non-locality, the fat graviton is able to "know" whether a loop diagram is a vacuum loop or a loop attached to external legs. In this way, the equivalence principle is preserved. However, it is not clear what detailed rules a fat graviton obeys. Also, the size of graviton is required as an input parameter. To solve the cosmological constant problem, the graviton size $l > 20$ microns seems to be a large number in the particle physics point of view.

There are also a number of scalar field models which can be viewed as modified gravity models. This is because the scalar part of gravity is coupled to matter. We shall review this class of models later, as part of the phenomenological models.

References

[1] H. A. Buchdahl, MNRAS **150** (1970) 1.
[2] A. Palatini, Rend. Circ. Mat. Palermo, **43** (1919) 203.
[3] A. Einstein, Sitzung-bet. Pruess. Akad. ICiss. **414** (1925) 37.
[4] M. Ferraris and M. Francaviglia, Gen. Rel. Grav. **14** (1982) 243.
[5] T. P. Sotiriou and V. Faraoni, Rev. Mod. Phys. **82** (2010) 451.
[6] B. Whitt, Phys. Lett. B **145** (1984) 176.
[7] J. D. Barrow and S. Cotsakis, Phys. Lett. B **214** (1988) 515.
[8] V. F. Mukhanov, H. A. Feldman, and R. H. Brandenberger, Phys. Rept. **215** (1992) 203.
[9] B. Huang, S. Li, and Y. G. Ma, Phys. Rev. D **81** (2010) 064003.
[10] S. M. Carroll *et al.*, Phys. Rev. D **70** (2004) 043528.
[11] T. Chiba, Phys. Lett. B **575** (2003) 1.
[12] S. M. Carroll *et al.*, New J. Phys. **8** (2006) 323.
[13] K. Bamba, C. Q. Geng, and S. Tsujikawa, Phys. Lett. B **688** (2010) 101; K. Bamba and C. Q. Geng, arXiv:1005.5234.
[14] M. Milgrom, ApJ. **270** (1983) 365.
[15] J. Bekenstein and M. Milgrom, ApJ. **286** (1984) 7.
[16] J. P. Bruneton, Phys. Rev. D **75** (2007) 085013.

[17] J. D. Bekenstein, in *Proceedings of the Second Canadian Conference on General Relativity and Relativistic Astrophysics*, ed. C. Dyer (World Scientific, Singapore, 1987).

[18] J. D. Bekenstein, in *Second Canadian Conference on General Relativity and Relativistic Astrophysics*, ed. A. Coley, C. Dyer and T. Tupper (World Scientific, Singapore, 1988).

[19] J. D. Bekenstein, in *Proceedings of the Sixth Marcel Grossmann Meeting on General Relativity*, ed. H. Sato and T. Nakamura (World Scientific, Singapore, 1992).

[20] J. D. Bekenstein, Phys. Rev. D **70** (2004) 083509.

[21] G. R. Dvali, G. Gabadadze, and M. Porrati, Phys. Lett. B **485** (2000) 208.

[22] E. Kiritsis *et al.*, JHEP **0302** (2003) 035; E. Kiritsis, JCAP **0510** (2005) 014; K. Umezu *et al.*, Phys. Rev. D **73** (2006) 063527.

[23] A. Nicolis, R. Rattazzi, and E. Trincherini, Phys. Rev. D **79** (2009) 064036.

[24] C. Brans and R. H. Dicke, Phys. Rev. **124** (1961) 925.

[25] L. Amendola, Phys. Rev. D **60** (1999) 043501.

[26] L. E. Qiang *et al.*, Phys. Rev. D **71** (2005) 061501(R); L. E. Qiang *et al.*, Phys. Lett. B **681** (2009) 210.

[27] B. Zwiebach, Phys. Lett. B **156** (1985) 315.

[28] S. Nojiri, S. D. Odintsov, and M. Sasaki, Phys. Rev. D **71** (2005) 123509.

[29] D. Lovelock, J. Math. Phys. **12** (1971) 498.

[30] M. H. Dehghani and S. Assyyaee, Phys. Lett. B **676** (2009) 16.

[31] P. Horava, Phys. Rev. D **79** (2009) 084008.

[32] E. N. Saridakis, Eur. Phys. J. C **67** (2010) 229.

[33] G. R. Bengochea and R. Ferraro, Phys. Rev. D **79** (2009) 124019.

[34] E. V. Linder, Phys. Rev. D **81** (2010) 127301.

[35] T. Harko *et al.*, arXiv:1104.2669.

[36] P. D. Mannheim, Prog. Part. Nucl. Phys. **56** (2006) 340.

[37] P. D. Mannheim and J. G. O'Brien, arXiv:1011.3495.

[38] P. D. Mannheim, arXiv:1005.5108.

[39] R. Sundrum, Phys. Rev. D **69** (2004) 044014.

10
Quantum Cosmology

As reviewed in Chapter 5, the approach of wave function of the universe either results in a probability $P(\Lambda) = \exp\left(\frac{3\pi}{G\Lambda}\right)$, or its exponential. The predicted cosmological constant is not only too small, but also conflicts with the thermal history of the universe. Several ways to solve the problem have been proposed.

For example, Firouzjahi, Sarangi and Tye [1, 2] brought the decoherence effect to the quantum creation of universe, and found the preferred classical universe is born at string scale, which is good for inflationary cosmology. As another approach, Hartle, Hawking and Hertog [3, 4] argued that from the Bayesian point of view, the quantum creation probability should multiply the spatial volume factor to get the observed probability. Again, inflation after the born of the universe is preferred.

These approaches on the one hand bring the Hartle–Hawking wave function back to be consistent with observations, but on the other hand make the whole subject more relevant to inflation, and less relevant to dark energy.

Nevertheless, dark energy is very probably an effect from quantum gravity. Indeed a lot of attempts on dark energy are related to quantum theories. Here in this chapter by quantum cosmology we shall restrict our attention to approaches closely related to the wave function of the universe. Other quantum approaches are reviewed in other related chapters.

10.1   Cosmological Constant Seesaw

The seesaw mechanism provides a connection between particles of large mass and small mass, which is extensively used in neutrino physics. Non-technically speaking, the (type I) seesaw mechanism relies on the fact that a matrix

$$M = \begin{pmatrix} 0 & x \\ x & y \end{pmatrix} \tag{10.1}$$

has eigenvalue

$$\lambda_\pm = \frac{y \pm \sqrt{y^2 + 4x^2}}{2}.$$

When $y \gg x$, the smaller eigenvalue is

$$\lambda_- \simeq -\frac{x^2}{y}, \tag{10.2}$$

which is suppressed by the larger matrix element y. Note that $\lambda_+ \lambda_- = -x^2$. Thus when adjusting y, one eigenvalue goes up, another goes down, like a seesaw. In neutrino physics, the analog of matrix M is the neutrino mass matrix.

The seesaw mechanism is aimed to explain why the neutrino mass is non-zero and small. This situation appears very similar to the cosmological constant problem. Based on this observation, Motl and Carroll proposed on their blogs [5, 6] the idea that the smallness of the cosmological constant may be explained by the seesaw mechanism as well. The observation is that the present cosmological constant, the Planck scale and the lowest possible SUSY scale have a hierarchy

$$M_\Lambda \sim M_{\text{SUSY}}^2/M_p, \tag{10.3}$$

where $M_\Lambda \equiv (\rho_\Lambda)^{1/4}$ is the energy scale of the dark energy. This equation is similar to (10.2).

This idea was realized using the wave function of the universe by McGuigan [7]. It is noticed that the cosmological constant term acts exactly like a mass term in the Wheeler–DeWitt equation. Consider a model with a set of massless free scalar fields ϕ^I minimally coupled to gravity, the homogeneous and isotropic background action can be written as

$$S = \frac{1}{2} \int dt a^3 N \left[-9M_P^2 \left(\frac{\dot{a}}{Na} \right)^2 + \frac{\dot{\phi}^I \dot{\phi}^I}{N^2} - \lambda + M_P^2 \frac{k}{a^2} \right], \tag{10.4}$$

where N is the lapse function. In Eq. (10.4) several coefficients are different from the literature. This can be understood as a different convention used in [7]. Write down the conjugate momentum and quantize the equation of motion, one gets the Wheeler–DeWitt equation

$$\left(-\frac{1}{M_P^2} \frac{\partial^2}{\partial V^2} + \frac{1}{V^2} \frac{\partial^2}{\partial \phi^{I2}} + M_P^2 \frac{k}{V^{2/3}} - \lambda \right) \Psi = 0, \tag{10.5}$$

where $V \equiv a^3$ is the volume of the universe, which can be understood as a time variable. The scalar fields ϕ^I, on the other hand, acts as spatial coordinates. Note that the term proportional to λ acts effectively as a mass term in this equation.

Up to now, everything comes as conventional. Now consider two universes with different cosmological constant and also with "coupling":

$$\left(-\frac{1}{M_P^2} \frac{\partial^2}{\partial V^2} + \frac{1}{V^2} \frac{\partial^2}{\phi^{I2}} + M_P^2 \frac{k}{V^{2/3}} - \lambda_1 \right) \Psi_1 + \sqrt{\lambda_1 \lambda_2} \Psi_2 = 0, \quad (10.6)$$

$$\left(-\frac{1}{M_P^2} \frac{\partial^2}{\partial V^2} + \frac{1}{V^2} \frac{\partial^2}{\phi^{I2}} + M_P^2 \frac{k}{V^{2/3}} - \lambda_1 - \lambda_2 \right) \Psi_2 + \sqrt{\lambda_1 \lambda_2} \Psi_1 = 0, \quad (10.7)$$

where the effective mass matrix now is

$$M^2 = \begin{pmatrix} 0 & \sqrt{\lambda_1} \\ \sqrt{\lambda_1} & \sqrt{\lambda_2} \end{pmatrix}^2. \quad (10.8)$$

This form is similar to Eq. (10.1). Now one can apply the seesaw mechanism, such that when

$$\lambda_1 \sim (10 \text{ TeV})^4, \qquad \lambda_2 \sim M_P^4, \quad (10.9)$$

the smaller eigenvalue becomes

$$\lambda_- \sim \frac{(10 \text{ TeV}^8)}{M_P^4}, \quad (10.10)$$

the correct order of magnitude for the current cosmological constant. However, it remains curious how the coupling constant is derived from first principle. In [8], Linde proposed that some interactions between universes could follow from averaging effects. It is interesting to see whether these interactions are of the type which we review in this section. Moreover, it is not clear in which form interacting universes will classically behave. It is clear that particles should stay in their mass eigenstates for energy and momentum measurement but the analog is not clear for quantum universes.

10.2   Wave Function through the Landscape

In quantum mechanics, there is a well-established mechanism of resonant tunneling, which is not only tested by experiments but also widely applied in the industry of electronics. Tye [9] applied the resonant tunneling mechanism to the string landscape, which provides a possible solution for the cosmological constant problem.

As a pure quantum effect, the probability for tunneling is typically exponentially suppressed. An intuitive understanding is that a bump in the potential, even higher than the energy of the quantum state, can not completely block the wave function. An exponentially small part of the wave function is leaked into the classically forbidden regime, which connects to another classically allowed regime.

However, things get changed when there are multiple bumps in the potential. As illustrated in Figure 10.1, when the incoming wave coming from region I has an energy equal to a bound state in region III, the probability for tunneling through the whole potential to region V is no longer exponentially suppressed by both of the barriers. Instead, the tunneling rate will be equal to the tunneling rate with the larger exponential suppression factor (of II or IV) divided by that with the smaller exponential suppression factor. Especially, when these two factors are of the same order, the tunneling rate becomes of order one.

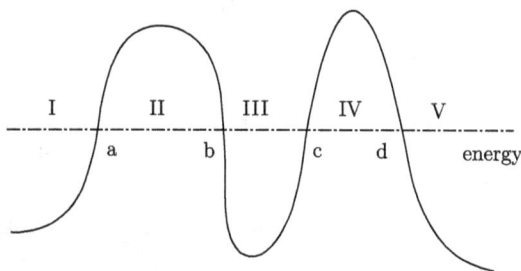

Figure 10.1   A resonant tunneling in quantum mechanics. The analog is applied to cosmology for the cosmological constant problem.

The reason for the odd thing to happen is as follows: In region V, the wave function only has outgoing component. In WKB approximation, the mode is $\sim e^{i \int pdq}$. This outgoing component can be linked to two modes in region IV, namely an exponentially decaying mode (here decay means becomes smaller from III to IV) $\sim e^{- \int pdq}$ and an exponentially growing mode $\sim e^{\int pdq}$. In conventional tunneling events without resonance, only the decaying mode is considered because the other one contributes an exponentially small part to the wave function in region III. Similarly, there are two modes in region II, only the decaying mode need to be considered for tunneling without resonance. However, when the incoming energy satisfies a bound state condition in region III,

$$\int pdq = \left(n + \frac{1}{2}\right)\pi, \qquad n = 0, 1, 2, \cdots, \tag{10.11}$$

the decaying mode in region IV can only link to the growing mode in region II, and the growing mode in IV can only link to the decaying mode in II, by requiring the wave function and its derivative are continuous at points a, b, c and d. Thus there must be one barrier in which the tunneling amplitude is suppressed and another barrier in which the amplitude is enhanced exponentially. This is why resonant tunneling can happen in quantum mechanics.

Further calculation shows that, for resonant tunneling to happen, the incoming energy should be exponentially close to Eq. (10.11). Otherwise the enhancement disappears. For a random incoming energy, the expectation value of tunneling coefficient $T(\text{I} \rightarrow \text{V})$ can be expressed as

$$T(\text{I} \rightarrow \text{V}) \sim \frac{T(\text{I} \rightarrow \text{III})T(\text{III} \rightarrow \text{V})}{T(\text{I} \rightarrow \text{III}) + T(\text{III} \rightarrow \text{V})}. \tag{10.12}$$

It is also straightforward to generalize the above two step resonant tunneling to n steps.

Tye [9] proposed that the universe may have experienced such resonant tunnelings in the string landscape. As reviewed in previous chapters, the string landscape is extremely complicated and in some sense random. Using Eq. (10.12), one can show that the tunneling probability is enhanced. Moreover, the string landscape is many-dimensional. The total tunneling rate Γ can be calculated by

$$\Gamma \sim n^d \Gamma_0, \tag{10.13}$$

where Γ_0 is the tunneling rate for a single tunneling event, d is the dimension of the landscape and n is the number of steps for resonant tunneling [10]. This is in sharp contrast with the case discarding resonance, where the tunneling rate is $\Gamma \sim (d \times \Gamma_0)^n$.

In this picture, the universe starts with a site in the landscape with large cosmological constant and tunnels through a series of sites to the state with the current cosmological constant, with reasonably large probability. It is assumed that the universe can not tunnel to any of the vacua with negative cosmological constant. The argument is that if the whole universe tunnels to a vacuum with negative cosmological constant, the entropy would decrease; while if a portion of the universe tunnels to such a vacuum, a singularity will develop.

The resonant tunneling picture gives an elegant understanding for the cosmological constant problem in the context of cosmic landscape, as long as one considers the mini-superspace model, where a closed universe as a whole could tunnel to other sites with resonance. In this case the quantum gravity problem reduces to a quantum mechanical problem. However, it is quite difficult to figure out what resonant tunneling looks like in quantum field theory. For example, in [11], it was proved that if the tunneling process, in terms of bubbles, satisfies a static boundary condition, there will be no such resonant tunneling in scalar quantum field theory. Later, it was shown in [12] that for contracting spherical bubbles, resonant tunneling is possible. Moreover, in [13], it was shown that the requirements in [11] are too restrictive. A most probable escape path does not necessarily satisfy the no-go theorem in [11], and a phase diagram is given to show that some of the potentials have resonant tunneling in scalar quantum field theory. Those resonant tunneling channels may be tested experimentally in the future by Helium experiments [14, 15]. For the case of gravity, and for the estimation of tunneling probabilities in a landscape, further investigations are deserved.

In [10], other rapid tunneling events were also considered, for example the Hawking–Moss tunneling without slow roll approximation. One reaches the conclusion that rapid tunneling can be common in the complicated string landscape thus the tunneling process become analogous to an electron in a random potential. Using renormalization group approach, Tye showed that after a number of successful tunneling events, the universe will settle down with localized wave function due to Anderson localization [16]. Similar ideas on Anderson localization of universe in a landscape were also discussed in [17] and [18]. Besides Anderson localization, decoherence of the quantum wave function may also eventually localize the universe to one site in the landscape [19].

References

[1] H. Firouzjahi, S. Sarangi, and S. H. H. Tye, JHEP **0409** (2004) 060.

[2] S. Sarangi and S. H. H. Tye, arXiv:hep-th/0505104.

[3] J. B. Hartle, S. W. Hawking, and T. Hertog, Phys. Rev. Lett. **100** (2008) 201301.

[4] S. W. Hawking, arXiv:0710.2029.

[5] http://motls.blogspot.com/2005/12/cosmological-constant-seesaw.html.

[6] http://blogs.discovermagazine.com/cosmicvariance/2005/12/05/duff-on-susskind/.

[7] M. McGuigan, arXiv:hep-th/0602112.

[8] A. D. Linde, arXiv:hep-th/0211048.

[9] S. H. H. Tye, arXiv:hep-th/0611148.

[10] S. H. H. Tye, arXiv:0708.4374.

[11] E. J. Copeland, A. Padilla, and P. M. Saffin, JHEP **0801** (2008) 066.

[12] P. M. Saffin, A. Padilla, and E. J. Copeland, JHEP **0809** (2008) 055.

[13] S. H. H. Tye and D. Wohns, arXiv:0910.1088.

[14] S. H. H. Tye and D. Wohns, arXiv:1106.3075.

[15] I. S. Yang, S. H. H. Tye and B.Shlaer, arXiv:1110.2045.

[16] P. W. Anderson, Phys. Rev. **109** (1958) 1492.

[17] D. Podolsky and K. Enqvist, JCAP **0902** (2009) 007.

[18] D. I. Podolsky, J. Majumder, and N. Jokela, JCAP **0805** (2008) 024.

[19] C. Kiefer, F. Queisser, and A. A. Starobinsky, arXiv:1010.5331.

11
Holographic Principle

11.1　The Holographic Principle

The essence of reality has been one of the central problems in philosophy and physics since thousands of years ago. Among all the discussions and debates on this subject, one of the most profound allegory is the Plato's Cave. In the recent decades, the inverse problem of Plato's Cave, under the name holographic principle [1, 2], has become one of the building blocks for a modern understanding of theoretical physics.

Holographic principle asserts that the world can be understood as a hologram. In other words a theory of gravity is dual to a boundary field theory without dynamical gravity in one less dimensions.

The story of holographic principle tracks back to the investigation of black hole physics. Starting from [3, 4], it is realized that classically, a stationary black hole in four dimensions is characterized by its mass, angular momentum and charge. A paradox arises here that matter that collapses into a classical black hole appears to lose almost all its entropy. This is a contradiction to the second law of thermodynamics. Moreover, Hawking proved that [5] the event horizon of a black hole never decreases with time. The paradox becomes sharper in this sense because it becomes hopeless for the "lost" entropy to come out.

The paradox was conceptually resolved by Bekenstein [6–8] by the conjecture that a black hole has an entropy proportional to its horizon area. The conjecture was soon proved by Hawking [9], with the form

$$S_{\mathrm{BH}} = \frac{A}{4G}\,.$$
(11.1)

Later, 't Hooft and Susskind [1, 2] realized that the black hole entropy can be understood as a dimensional reduction, or holographic principle.

The entropy is viewed as degrees of freedom measured in Planck units, which lives on the surface of the strongly gravitating system.

The right hand side of Eq. (11.1) is not only the entropy of a black hole, but also an entropy bound for any form of matter localized in a spherical region

$$S_{\text{matter}} \leqslant \frac{A}{4G}. \tag{11.2}$$

For example, consider a spherical symmetric shell of photons which fall towards the center. Before the photons form a black hole, the left hand side of Eq. (11.2) is always smaller than the right hand side. When the photons hit the Schwarzschild radius, the bound may be saturated when the photons have maximal possible entropy, in which case the wavelength of the photon equals to the Schwarzschild radius.

However, as pointed out by Cohen, Kaplan and Nelson [10], the thermal entropy of an effective field theory can never saturate the entropy bound (11.2). This is because mass M and entropy S of a field theory scale with the length scale of the system (infrared cutoff) and the temperature as

$$M \sim L^3 T^4, \qquad S \sim L^3 T^3. \tag{11.3}$$

At the point just before the formation of a black hole, the field theory has Schwarzschild mass, however the entropy takes the form

$$S \sim (M_{\text{P}} L)^{3/2} \sim S_{\text{BH}}^{3/4}, \tag{11.4}$$

which is always smaller than the black hole entropy. Thus, for thermal field theory matter, Schwarzschild mass behaves as a tighter bound than the entropy bound. In other words, let the ultraviolet cutoff of the system be Λ_{UV}, the maximum energy density in the effective field theory $\rho \sim \Lambda_{\text{UV}}^4$ must satisfy

$$L^3 \Lambda_{\text{UV}}^4 \sim E \leqslant L M_{\text{P}}^2. \tag{11.5}$$

Equation (11.4) could also be written as follows: the maximal allowed energy density ρ satisfies

$$\rho = 3\mathbf{c}^2 M_{\text{P}}^2 L^{-2}, \tag{11.6}$$

where $\mathbf{c}$ is a number introduced in [11].

11.2 Holographic Dark Energy

The reasoning of last section can be applied to the vacuum, which leads to a holographic model of dark energy. The question is how to choose the infrared cutoff. As pointed out by Hsu [12], the simplest choice $L = 1/H$ does not work because it has a wrong equation of state. Li [11] pointed out that if one take $L = 1/R_h$, where R_h is the future event horizon defined as

$$R_h = a \int_t^\infty \frac{dt}{a} = a \int_a^\infty \frac{da}{Ha^2}, \qquad (11.7)$$

the energy density (11.6) becomes

$$\rho_{de} = 3c^2 M_P^2 R_h^2, \qquad (11.8)$$

which does behave as dark energy, with the Friedmann equation

$$3M_P^2 H^2 = \rho_{de} + \rho_m + \cdots, \qquad (11.9)$$

where $\rho_m + \cdots$ denotes matter and other components in the universe.

To see this, note the index of equation of state w_{de} can be defined as

$$\rho'_{de} + 3(1 + w_{de})\rho_{de} = 0, \qquad (11.10)$$

where prime denotes derivative with respect to $\ln a$. Take derivative of Eq. (11.8) and use Eq. (11.10) to substitute ρ'_{de}, one can get

$$w_{de} = -\frac{1}{3} - \frac{2}{3HR_h} = -\frac{1}{3} - \frac{2\sqrt{\Omega_{de}}}{3c}. \qquad (11.11)$$

This equation is about the nature of holographic dark energy itself, which is independent of what form of matter is present in the universe. Here Ω_{de} is the relative energy density of holographic dark energy, defined as

$$\Omega_{de} \equiv \frac{\rho_{de}}{3M_P^2 H^2} = \frac{c^2}{R_h^2 H^2}. \qquad (11.12)$$

Ω_{de} turns out to be a convenient variable for solving equations of motion for holographic dark energy.

Before going to the equation of motion, now one can already find out qualitative behavior of holographic dark energy from Eq. (11.12). When the holographic dark energy is sub-dominant ($\Omega_{de} \ll 1$), $w_{de} \simeq -1/3$ thus $\Omega_{de} \sim a^{-2}$. When the holographic dark energy is dominant ($\Omega_{de} \simeq 1$),

$w_{de} \simeq -1/3 - 2/3\mathbf{c}$, thus the universe experiences accelerating expansion as long as $\mathbf{c} > 0$.

Now consider the universe dominated by holographic dark energy and pressureless matter. Taking derivative of Eq. (11.12) with respect to $\ln a$, and making use of the Friedmann equation (11.9), one gets an equation for Ω_{de} as

$$\frac{\Omega'_{de}}{\Omega_{de}} = (1 - \Omega_{de})\left(1 + \frac{2\sqrt{\Omega_{de}}}{\mathbf{c}}\right). \qquad (11.13)$$

The equation can be solved as

$$\ln(a/a_0) = \ln \Omega_{de} + \frac{\mathbf{c}\ln(1 + \sqrt{\Omega_{de}})}{2 - \mathbf{c}} - \frac{\mathbf{c}\ln(1 - \sqrt{\Omega_{de}})}{2 + \mathbf{c}}$$

$$- \frac{8\ln(\mathbf{c} + 2\sqrt{\Omega_{de}})}{4 - \mathbf{c}^2}, \qquad (11.14)$$

where a_0 is a constant. The phenomenological implication of this solution will be discussed in later chapters.

It was also noticed in [11] that during inflation holographic dark energy is diluted. To have the correct fraction of dark energy at present time, one requires about 60 e-folds of inflation. In other words, holographic dark energy provides an explanation of the coincidence problem, as long as inflation only lasts for about 60 e-folds, before which dark energy also dominates. The detailed implication for inflation from holographic dark energy was considered in [13].

As an energy component with negative pressure, one might question the stability of holographic dark energy against fluctuations [14]. However, the dynamics of holographic dark energy is actually not governed by that of a perfect fluid. Instead it is governed by the dynamics of the future event horizon. The fluctuation of the future event horizon can be written as

$$\delta\rho_{de} = -2\rho_{de}\frac{\delta R_h}{R_h}. \qquad (11.15)$$

This fluctuation can be analyzed using cosmic perturbation theory. Spherical symmetric perturbations were calculated in [15], and it was shown that the perturbation approaches a constant at late times thus the background is stable against the fluctuations.

Phenomenologically it is interesting to investigate interactions between holographic dark energy and matter components [16]. The interaction can be added to the continuity equation as

$$\rho'_{de} + 3(1 + w_{de})\rho_{de} = 3b\rho_i \,, \tag{11.16}$$

where ρ_i can be set to dark energy density ρ_{de}, matter density ρ_m, critical energy density ρ_c, or their combinations.

Finally, one should note that holographic dark energy solves both the old and new cosmological problem in a consistent way. This is unlike a number of other models, in which one has to assume the solution of the old cosmological constant problem, i.e. assume the vacuum energy to be zero, and propose a small dark energy on top of that.

11.3   Complementary Motivations

Besides motivation from [10], there are also a number of other theoretical motivations leading to the form of holographic dark energy, among which some are motivated by holography and others from other principles of physics. We shall briefly review some of the motivations in this section.

• Casimir energy in de Sitter space.

The Casimir energy of electromagnetic field in static de Sitter space was calculated in [17, 18]. The Casimir energy can be written as

$$E_{\text{Casimir}} = \frac{1}{2} \sum_\omega |\omega|, \tag{11.17}$$

where the absolute value of ω is the energy with respect of the time of the static patch. E_{Casimir} can be calculated using heat kernel method with ζ function regularization. The result is

$$E_{\text{Casimir}} = \frac{3}{8\pi} \left(\ln \mu^2 - \gamma - \frac{\Gamma'(-1/2)}{\Gamma(-1/2)} \right) \left(\frac{L}{l_P^2} - \frac{1}{L} \ln \left(\frac{2L}{l_P^2} \right) \right)$$
$$+ \mathcal{O}(1/L) \,, \tag{11.18}$$

where L is the de Sitter radius, γ is the Euler constant and $\Gamma'(-1/2) \simeq -3.48$. Here a cutoff at stretched horizon is imposed, which has a distance l_P away from the classical horizon. Note that the dominate term scales as $E_{\text{Casimir}} \sim L/l_P^2$. Thus the energy density scales as $\rho_{\text{Casimir}} \sim M_P^2 L^{-2}$, which is the form of holographic dark energy.

• Quantum uncertainty of transverse position.

At Planck scale, gravity becomes strongly coupled and the classical spacetime picture breaks down. It was suggested by Hogan [19, 20] that the Planck scale quantum gravitational effect could be modeled by a random noise. A particle moving a distance l_P will have a kick of the same order l_P in the transverse direction. As the kick in the transverse direction is random, the summation of n kicks results in a random walk of distance $\sqrt{n}l_P$. Thus when a particle moves distance L, there is an uncertainty in the transverse direction of order

$$\Delta X = \sqrt{Ll_P} \, . \tag{11.19}$$

This relation looks quite like the energy bound (11.6). Indeed one can show that these two bounds are related. When one take L to be the scale of the universe, ΔX as an infrared cutoff gives the energy density for holographic dark energy.

• Entanglement entropy from quantum information theory.

The vacuum entanglement energy was considered in the cosmological context by [21]. The entanglement entropy of the quantum field theory vacuum with a horizon can be generically written as

$$S_{\mathrm{Ent}} = \frac{\beta R_{\mathrm{h}}^2}{l^2} \, , \tag{11.20}$$

where β is an order one constant and l is the ultraviolet cutoff from quantum gravity. The entanglement energy is conjectured to satisfy

$$dE_{\mathrm{Ent}} = T_{\mathrm{Ent}} dS_{\mathrm{Ent}} \, , \tag{11.21}$$

where $T_{\mathrm{Ent}} = 1/(2\pi R_{\mathrm{h}})$ is the Gibbons–Hawking temperature. Integrate Eq. (11.21), one gets

$$E_{\mathrm{Ent}} = \frac{\beta N_{\mathrm{dof}} R_{\mathrm{h}}}{\pi l^2} \, , \tag{11.22}$$

where N_{dof} is the number of light fields present in the vacuum. Thus the energy density is

$$\rho_{\mathrm{de}} = 3\mathbf{c}^2 M_P^2 R_{\mathrm{h}}^{-2} \, , \qquad \mathbf{c} = \frac{\sqrt{\beta N_{\mathrm{dof}}}}{2\pi l M_P} \, . \tag{11.23}$$

Here $\mathbf{c}$ is in principle calculable in the quantum information theory.

• Dark energy from entropic force.

Verlinde conjectured [22] that gravity may be an entropic force, instead of a fundamental force of nature. [23] investigated the implication of the conjecture for dark energy. It is suggested that the entropy change of the future event horizon should be considered together with the entropy change of the test holographic screen. Consider a test particle with physical radial coordinate R, which is the distance between the particle and the "center" of the universe where the observer is located. The energy of the future event horizon, using Verlinde's proposal, can be estimated as

$$E_h \sim N_h T_h \sim R_h/G\,, \qquad (11.24)$$

where $N_h \sim R_h^2/G$ is the number of degrees of freedom on the horizon, and $T_h \sim 1/R_h$ is the Gibbons–Hawking temperature. Following Verlinde's argument (instead of Newtonian mechanics), the energy of the horizon induces a force to a test particle of order $F_h \sim GE_h m/R^2$, which can be integrated to obtain a potential

$$V_h \sim -\frac{R_h m}{R} = -\mathbf{c}^2 m/2\,, \qquad (11.25)$$

where after the integration one can take the limit $R \to R_h$, and c is a constant reflecting the order one arbitrarily. Using standard argument leading to Newtonian cosmology, this potential term for a test particle will show up in the Friedmann equation as a component of dark energy $\rho_{de} = 3\mathbf{c}^2 M_P^2 R_h^{-2}$. Again it is the form of holographic dark energy.

• Holographic gas as dark energy.

The nature of a general strongly correlated gravitational system is not well understood. In [24] it was suggested that the quasi-particle excitations of such a system may be described by holographic gas, with modified degeneracy

$$\omega = \omega_0 k^a V^b M_P^{3b-a}\,, \qquad (11.26)$$

where ω_0 is a dimensionless constant. Inspired by holography, when taking $T \propto V^{-1/3}$ and $S \propto V^{2/3}$, one needs $b = (a+2)/3$ and the energy density ρ can be written as

$$\rho_{de} = \frac{a+3}{a+4}\frac{ST}{V}\,, \qquad (11.27)$$

where S, T, V are the entropy, temperature and volume of the system. Applying to cosmology, $S = 8\pi^2 R^2 M_P^2$ and $T = 1/(2\pi R)$, one obtains

$$\rho_{de} = 3\frac{a+3}{a+4}M_P^2 R^{-2}\,. \qquad (11.28)$$

This has the same form as holographic dark energy with

$$\mathbf{c}^2 = \frac{a+3}{a+4} \, . \tag{11.29}$$

There are some other alternative motivations for holographic dark energy. For example, the relation between holographic dark energy and vacuum decay was discussed in [25]. In addition, there are also many extended versions of holographic dark energy, see [26] and references therein for more details.

11.4   Agegraphic Dark Energy

In this section we review another dark energy model motivated from holographic physics, named agegraphic dark energy by Cai [27] and later Wei and Cai [28]. As discussed in [27, 28], there is a subtlety in the original version of agegraphic dark energy model [27], where cosmic time is used as the age cutoff. Thus here we mainly review the so-called new agegraphic dark energy model [28].

The agegraphic dark energy model is based on the Karolyhazy [29] uncertainty principle

$$\delta t = \beta t_{\mathrm{P}}^{2/3} t^{1/3} \, , \tag{11.30}$$

where β is an order one constant and t_{P} is Planck time. It was noticed in [27] and [30] that this relation has close relation with holographic principle and black hole entropy bound. Based on the Karolyhazy relation, Maziashvili derived an energy density of the vacuum energy

$$\rho_{\mathrm{de}} = \frac{3n^2 m_{\mathrm{P}}^2}{t^2} \, , \tag{11.31}$$

where n is a numerical factor as introduced in [27]. Wei and Cai proposed that, when the time in the above formula takes the form of conformal time

$$\eta = \int \frac{dt}{a} = \int \frac{da}{a^2 H} \, , \tag{11.32}$$

the energy density (11.31) can be well behaved as a dark energy component.

The equation of motion for Ω_{de} takes the form

$$\frac{d\Omega_{\mathrm{de}}}{da} = \frac{\Omega_{\mathrm{de}}}{a}(1 - \Omega_{\mathrm{de}})\left(1 - \frac{2}{n}\frac{\sqrt{\Omega_{\mathrm{de}}}}{a}\right) \, . \tag{11.33}$$

Note that the scale factor a appears explicitly.

The index of equation of state for agegraphic dark energy takes the form

$$w_{\rm de} = -1 + \frac{2}{3n} \frac{\sqrt{\Omega_{\rm de}}}{a}, \tag{11.34}$$

thus there can be accelerating solutions. Especially at late times when $a \to \infty$, $w_{\rm de} \to -1$.

Unlike the case of holographic dark energy, here $\Omega_{\rm de}$, n and a should satisfy an additional consistency relation. For example, in the matter dominated universe the conformal time $\eta \propto a^{1/2}$, thus $\rho_{\rm de} \propto a^{-1}$. Comparing with Eq. (11.34), one obtains $\Omega_{\rm de} = n^2 a^2/4$. Similarly, in radiation dominated universe the corresponding relation is $\Omega_{\rm de} = n^2 a^2$.

With the consistency relation, agegraphic dark energy is a model with a single parameter. This is different from holographic dark energy where there are two parameters c and $\Omega_{\rm m0}$.

11.5   Ricci Dark Energy

Another natural choice for cosmological infrared cutoff is the intrinsic curvature of the universe. Based on this, Gao, Chen and Shen [31] proposed a model of Ricci dark energy.[1] In the Ricci dark energy model, the energy density is

$$\rho_{\rm de} = \frac{3\alpha}{8\pi} \left(\dot{H} + 2H^2 + \frac{k}{a^2} \right) = -\frac{\alpha}{16\pi} \mathcal{R}, \tag{11.35}$$

where $\mathcal{R}$ is the Ricci scalar.

With the proposed form of energy density, the energy density can be solved from the Friedmann equation as

$$\rho_{\rm de} = \frac{\alpha}{2 - \alpha} \Omega_{\rm m0} e^{-3 \ln a} + f_0 e^{-\left(4 - \frac{\alpha}{2}\right) \ln a}, \tag{11.36}$$

where f_0 is an integration constant. The pressure can be solved from energy conservation,

$$p_{\rm de} = -\left(\frac{2}{3\alpha} - \frac{1}{3} \right) f_0 e^{-\left(4 - \frac{2}{\alpha}\right) \ln a}. \tag{11.37}$$

The model parameter α and f_0 are to be determined by data fittings. For some extended versions of Ricci dark energy, see [33–35].

---

[1] In [32] Nojiri and Odintsov proposed a generalized framework of holographic dark energy, which contains Ricci dark energy as a special case.

References

[1] G. 't Hooft, arXiv:gr-qc/9310026.

[2] L. Susskind, J. Math. Phys. **36** (1995) 6377.

[3] W. Israel, Phys. Rev. **164** (1967) 1776.

[4] W. Israel, Commun. Math. Phys. **8** (1968) 245.

[5] S. W. Hawking, Phys. Rev. Lett. **26** (1971) 1344.

[6] J. D. Bekenstein, Lett. Nuovo Cim. **4** (1972) 737.

[7] J. D. Bekenstein, Phys. Rev. D **7** (1973) 2333.

[8] J. D. Bekenstein, Phys. Rev. D **9** (1974) 3292.

[9] S. W. Hawking, Nature **248** (1974) 30.

[10] A. G. Cohen, D. B. Kaplan, and A. E. Nelson, Phys. Rev. Lett. **82** (1999) 4971.

[11] M. Li, Phys. Lett. B **603** (2004) 1.

[12] S. D. H. Hsu, Phys. Lett. B **594** (2004) 13.

[13] B. Chen, M. Li, and Y. Wang, Nucl. Phys. B **774** (2007) 256.

[14] Y. S. Myung, Phys. Lett. B **652** (2007) 223.

[15] M. Li, C. S. Lin, and Y. Wang, JCAP **0805** (2008) 023.

[16] B. Wang, Y. G. Gong, and E. Abdalla, Phys. Lett. B **624** (2005) 141.

[17] M. Li, R. X. Miao, and Y. Pang, Phys. Lett. B **689** (2010) 55.

[18] M. Li, R. X. Miao, and Y. Pang, Opt. Express **18** (2010) 9026.

[19] C. J. Hogan, arXiv:astro-ph/0703775.

[20] C. J. Hogan, arXiv:0706.1999.

[21] J. W. Lee, J. Lee, and H. C. Kim, JCAP **0708** (2007) 005.

[22] E. P. Verlinde, arXiv:1001.0785.

[23] M. Li and Y. Wang, Phys. Lett. B **687** (2010) 243.

[24] M. Li *et al.*, Commun. Theor. Phys. **51** (2009) 181.

[25] E. Elizalde *et al.*, Phys. Rev. D **71** (2005) 103504.

[26] Y. G. Gong, Phys. Rev. D **70** (2004) 064029; H. Wei, Commun. Theor. Phys. **52** (2009) 743; K. Karami and J. Fehri, Phys. Lett. B **684** (2010) 61; F. Adabi *et al.*, arXiv:1105.1008.

[27] R. G. Cai, Phys. Lett. B **657** (2007) 228.

[28] H. Wei and R. G. Cai, Phys. Lett. B **660** (2008) 113.

[29] F. Karolyhazy, Nuovo Cim. A **42** (1966) 390.

[30] M. Maziashvili, Phys. Lett. B **652** (2007) 165.

[31] C. J. Gao *et al.*, Phys. Rev. D **79** (2009) 043511.

[32] S. Nojiri and S. D. Odintsov, Gen. Rel. Grav. **38** (2006) 1285.

[33] L. N. Granda and A. Oliveros, Phys. Lett. B **671** (2009) 199.

[34] L. X. Xu and Y. T. Wang, JCAP **06** (2010) 002; Y. T. Wang and L. X. Xu, Phys. Rev. D **81** (2010) 083523.

[35] L. P. Chimento, M. Forte and M. G. Richarte, arXiv:1106.0781.

12
Back-Reaction

A universe without nonlinearity is simple and simply dull. Physicists like to start their calculation from the linear case because it is mathematically easy and often solvable. However, nobody likes to live in such a universe where two waves always propagate through each other without any impact.

In modern cosmology, one of the most important assumption is the cosmological principle, which states that the universe is homogeneous and isotropic on (observably) large scales. In the standard setup, the cosmological principle is encoded into the FRW metric and all the subsequent conclusions, especially the Friedmann equation, are under this assumption.[1]

The cosmological principle is indeed supported strongly by cosmological experiments on CMB and large scale structure (LSS). However, at late times and on small scales (sub-Hubble scales), the universe is not homogeneous at all. There are all kinds of structures in the universe. On the other hand, on scales larger than the observable universe (super-Hubble scales), it is not known whether the universe remains homogeneous and isotropic or not. Due to the nonlinearity of gravity, these sub-Hubble or super-Hubble scales may back-react on to the scale of the observable universe.

As early as in 1931, Einstein [1] already mentioned that the matter distribution is in reality inhomogeneous and the approximate treatment (cosmological principle) may be illusionary, when he was trying to explain why the Hubble's value of the Hubble parameter is about ten times too large [2]. Half a century later, Ellis re-examined the effect of clumpiness on

---

[1] The Friedmann equation, coupled to the matter equation of motion, becomes already non-linear at the background level. While the non-linearity we are discussing here is on the perturbation level. By definition, linear perturbation will not back-react the background. Back-reaction from perturbations is possible to show up only when nonlinear fluctuations are considered.

the average, under the name fitting problem [3, 4]. This started the modern story of back-reaction as an effective component of dark energy.

12.1 Sub-Hubble Inhomogeneities

In this section, we consider back-reaction from sub-Hubble scale fluctuations. Typically, the aim of the sub-Hubble scale back-reaction theory is assuming there is no cosmological constant (i.e. the old cosmological constant problem is solved), and providing cosmological acceleration from back-reaction of sub-Hubble scale inhomogeneities. Most details of this section can be found in [5]. We assume in this section that the universe is dominated by pressureless dust.

To consider small scale fluctuations, it is useful to derive local versions of the Friedmann equation and the continuity equation. To do this, it is helpful to decompose the derivative of the velocity field into components, where each component has clear meaning in the sense of fluid mechanics:

$$u_{\mu;\nu} = \frac{1}{3}\theta h_{\mu\nu} + \sigma_{\mu\nu} + \omega_{\mu\nu}\,, \tag{12.1}$$

where u^μ is the four velocity of the fluid, and $h_{\mu\nu} = g_{\mu\nu} - u_\mu u_\nu$ is the spatial projection of the metric.

On the right hand side of Eq. (12.1), the scalar part θ is called volume expansion scalar, defined as $\theta \equiv u^\mu{}_{;\mu}$, which measures the local expansion of the fluid. In the familiar homogeneous and isotropic FRW universe, θ is reduced to $3H$. The symmetric part $\sigma_{\mu\nu}$ is defined by $\sigma_{\mu\nu} = u_{(\mu;\nu)} - \theta h_{\mu\nu}/3$ and describes the shear of the fluid. The anti-symmetric part $\omega_{\mu\nu}$ is defined by $\omega_{\mu\nu} \equiv u_{[\mu;\nu]}$ and describes the vorticity.

Taking one more covariant derivative on Eq. (12.1), using the commutative relation for covariant derivative to relate the derivatives to curvature, and using the Einstein equation to relate the curvature to stress tensor, one can obtain

$$\dot{\theta} + \frac{1}{3}\theta^2 = -4\pi G\rho - 2\sigma^2 + 2\omega^2\,. \tag{12.2}$$

This equation is known as the Raychaudhuri equation [6], which is widely used in general relativity. Another useful equation from the combination of the Einstein equation and the decomposition equation (12.1) is

$$\frac{1}{3}\theta^2 = 8\pi G\rho - \frac{1}{2}R_3 + \sigma^2 - \omega^2\,, \tag{12.3}$$

which is the local version of the Friedmann equation, where R_3 is the spatial curvature on the slice orthogonal to u^μ, $\sigma^2 \equiv \sigma^{\mu\nu}\sigma_{\mu\nu}/2$, and $\omega^2 \equiv \omega^{\mu\nu}\omega_{\mu\nu}/2$. Finally the continuous equation for matter component takes the form

$$\dot{\rho} + \theta\rho = 0\,. \tag{12.4}$$

One can also derive equations for the time evolution of $\sigma_{\mu\nu}$ and $\omega_{\mu\nu}$ from the Einstein equation, but we will not need them here.

Now we are about to average these local analog of the Friedmann equations. The spatial average operation is defined as

$$\langle f \rangle(t) \equiv \frac{\int d^3x \sqrt{g_3(t,x)} f(t,x)}{\int d^3x \sqrt{g_3(t,x)}}\,, \tag{12.5}$$

where g_3 is the determinant of the three dimensional metric.

Using this definition, the local equations (12.2), (12.3) and (12.4) can be written as

$$3\frac{\ddot{a}}{a} = -4\pi G \langle \rho \rangle + \mathcal{Q}\,, \tag{12.6}$$

$$3H^2 = 8\pi G \langle \rho \rangle - \frac{1}{2}\langle R_3 \rangle - \frac{1}{2}\mathcal{Q}\,, \tag{12.7}$$

$$\partial_t \langle \rho \rangle + 3H\langle \rho \rangle = 0\,. \tag{12.8}$$

Equations (12.6), (12.7) and (12.8) are called the Buchert equations [7], where the averaged scale factor a is defined as

$$a(t) \equiv \left(\frac{\int d^3x \sqrt{g_3(t,x)}}{\int d^3x \sqrt{g_3(t_0,x)}} \right)^{1/3}\,, \qquad H \equiv \frac{\dot{a}}{a}\,. \tag{12.9}$$

Note that a and H used in this section denote averaged variables, not to confuse with those in other sections. $\mathcal{Q}$ is defined as

$$\mathcal{Q} \equiv \frac{2}{3}\left(\langle \theta^2 \rangle - \langle \theta \rangle^2 \right) - 2\langle \sigma^2 \rangle\,. \tag{12.10}$$

In the Buchert equations (12.6) and (12.7), the back-reaction variable $\mathcal{Q}$ is the novel term compared from the familiar FRW equations. $\mathcal{Q}$ can be thought of emergent in the sense of coarse graining. When $\mathcal{Q} > 0$, $\mathcal{Q}$ will behave as an effective component in the universe, which drives the late time acceleration.

So far we have not said anything about how to choose spatial slices to do the average. In fact choosing spatial slices properly is extremely important for the back-reaction calculation to make correct predictions, both theoretically and phenomenologically.

Theoretically, in general relativity there is no preferred choice for the spatial slices. In linear cosmic perturbation theory, there is an elegant gauge invariant way to do calculation. However, here the back-reaction one considers is beyond linear order and a gauge invariant formalism is not available on sub-Hubble scales.

Phenomenologically, cosmological experiments are typically carried out by measuring redshift and distance. When the universe is perturbed one has to make sure whether the calculated scale factor is indeed the one which is used to calculate these quantities.

For example, as pointed out by Ishibashi and Wald [8], if one averages two disconnected decelerating universes, one could get the conclusion that the "whole" universe is accelerating. This absurd conclusion shows that great care is needed to select the averaging hypersurface. For this purpose, one is led to consider the light propagation in a perturbed space.

In [9], Rasanen showed that the redshift in a dusty universe can be calculated as

$$1 + z = \exp\left\{ \int_{\eta}^{\eta_0} d\eta \left(\frac{1}{3}\theta + \sigma_{\mu\nu}e^{\mu}e^{\nu} \right) \right\}, \qquad (12.11)$$

where the integral is along the null geodesic, and e^{μ} is along the spatial direction of the geodesic. One can argue that when there is no preferred directions in the sky, the integration of $\sigma_{\mu\nu}e^{\mu}e^{\nu}$ should be suppressed by averaging effect, as long as one choose the spatial slice with statistical homogeneity and isotropy.

Similarly, it was shown in [10] that for angular diameter distance and the luminosity distance, similar average effects occur for statistical homogeneous and isotropic slicing. These results show evidence that the statistical homogeneous and isotropic slicing is the one that should be used when the calculation of back-reaction need to be compared with observations.

Having determined the choice of slicing, one can focus on the calculation of the back-reaction variable $\mathcal{Q}$. Unfortunately the calculation is very difficult. The Newtonian calculation might be oversimplified, as in Newtonian gravity the two terms in Q cancels up to a surface term [11]. For calculations in general relativity, only toy models are doable currently. Some calculations report there is strong cancellation between the two terms in Q, although they do not exactly cancel [12]. Others showed that it is not the case [13]. Thus whether or not small scale back-reaction could become important deserves further investigation.

Besides the approach of averaging, there is also a Lemaître–Tolman–Bondi [14–16] approach of sub-Hubble back-reaction. The idea is that there are voids in the large scale structure. We may live in a void in which the observables such as the luminosity distance need to be calculated with more care. As an example, in [17, 18], it was shown that the LTB type back-reaction could mimic dark energy.

12.2   Super-Hubble Inhomogeneities

Compared with the sub-Hubble theories, the theories of super-Hubble back-reaction has a wider variety of goals. Some of the works aim to provide a screening mechanism for the cosmological constant, and completely solve the cosmological constant problem. While some other works, like sub-Hubble theories, aims to give acceleration assuming the cosmological constant is zero. The difference is that, screening mechanism gives a negative energy density contribution from back-reaction, while the latter gives a positive energy density from back-reaction.

For the screening mechanism, Mukhanov, Abramo and Brandenberger [19, 20] have set up a gauge invariant formalism for perturbations up to second order. The quantities they use are gauge invariant after spatial integration (averaging). For scalar field matter $\varphi = \varphi_0 + \delta\varphi$, the second order perturbated stress tensor is calculated. Interestingly, the second order stress tensor, in the slow roll approximation, has an equation of state $p = -\rho > 0$, in other words, it is a cancellation or screening of cosmological constant. There are also similar results from gravitational loop calculations by earlier studies of Tsamis and Woodard [21, 22].

However, Unruh [23] (see also [24]) pointed out that gauge invariance of spatial averaged variables does not guarantee that the calculated effect is accessible or observable by a local observer. This objection was supported by Geshnizjani and Brandenberger [25] from an explicit calculation of a scalar field coupled to gravity. It is shown that without isocurvature fluctuations, the expansion θ, which is locally accessible, do not receive any back-reactions from super-Hubble perturbations. The expansion θ including back-reactions, as a function of $\varphi = \varphi_0 + \delta\varphi$, has the same function dependence as the background expansion θ_0, as a function of φ_0:

$$\theta = \sqrt{3GV(\phi)}\,. \qquad (12.12)$$

Thus for super-Hubble perturbations, a local observer will observe φ and can not find a difference in dynamics compared with the background.

However, with isocurvature perturbations, the situation changes. In [26], Geshnizjani and Brandenberger considered back-reaction in two scalar field model and found that in this case the back-reaction does not vanish. The reason is that there are different choices for local clocks. For different choice of clocks (proper time clock and energy density for a scalar field, as considered in [26]), the results of back-reaction are different (actually the correction changes sign). Thus it is crucial to identify which is the observable related to observations of dark energy.

Abramo and Woodard [27, 28] proposed a local operator (before spatial average) and showed that the back-reaction does not vanish. Further calculation [29–33] showed from one-loop and two-loop calculation that the probe scalar field obtains an equation of state $p < -\rho$, which may drive a period of super-acceleration. Eventually the super-acceleration will be turned off by the non-zero renormalized mass. Based on these consideration, a scenario of non-local cosmology was proposed to model the above behavior [34–36] (see also [37] and references therein).

The back-reaction from scalar field to de Sitter space as a screening mechanism was also considered in [38–40]. The authors argued that the de Sitter space analytically continued from a sphere should not be used to describe realistic cosmology because it corresponds to de Sitter space artificially kept at fixed Gibbons–Hawking temperature, which corresponds to de Sitter space surrounded by reflecting walls. Instead, They consider a period of de Sitter expansion

$$ds^2 = -dt^2 + a(t)^2 d\vec{x}^2, \qquad a(t) = e^{T\tanh(t/T)}, \qquad (12.13)$$

where $T \gg 1/H$ is a time cutoff such that interactions from de Sitter background are removed in the asymptotic past and future. Correspondingly, the vacuum choice is that the field approaches local Minkowski vacuum at $t \to -\infty$, which is different from the Bunch–Davies vacuum.

The calculations in [40] showed results as follows. For free scalar fields, as a result of de Sitter symmetry, the correction to the one point function is a trivial redefinition of the cosmological constant. For interacting fields in the FRW patch of the de Sitter space, two point function of the scalar correlator produces nontrivial infrared divergence but the one point function is not affected. In global de Sitter space, where the accelerating expansion of the universe is initialized by a period of contraction, there are nontrivial effects for the one point function for scalar fields. However, for the case of dark energy, our universe is clearly not contracting right before dark energy domination. Thus it remains an open question whether this approach could provide a dynamical explanation for the dark energy problem. Nevertheless, the interaction which breaks the de Sitter symmetry is an interesting candidate for dynamical dark energy.

Some other screening mechanisms are also reviewed in the previous chapter on tuning mechanisms, where the issue of back-reaction is not as relevant as discussed here.

As mentioned above, there is another class of super-Hubble back-reaction models, which produce a cosmological constant instead of screening it. For example, Kolb, Matarrese, Notari and Riotto considered the effect of super-Hubble perturbations from inflation. These perturbations, viewed from a local observer, may look like a source of acceleration. There are also counter arguments on these class of mechanisms, see for example [41].

One should also note that there is a whole literature for loop calculation during inflation, which is another accelerating epoch of the universe. The techniques developed in those loop calculations are also helpful in understanding the dynamics of dark energy. But we shall not introduce these works here. Interested readers are referred to [42] and related works.

References

[1] A. Einstein, Sitz. Preuss. Akad. Wiss. Phys-Math **235** (1931) 37.
[2] N. Straumann, arXiv:gr-qc/0208027.
[3] G. F. R. Ellis, Gen. Rel. Grav. **41** (2009) 581.
[4] G. F. R. Ellis and W. Stoeger, Class. Quant. Grav. **4** (1987) 1697.
[5] S. Rasanen, arXiv:1102.0408.
[6] A. Raychaudhuri, Phys. Rev. **98** (1955) 1123.
[7] T. Buchert, Gen. Rel. Grav. **32** (2000) 105.
[8] A. Ishibashi and R. M. Wald, Class. Quant. Grav. **23** (2006) 235.
[9] S. Rasanen, JCAP **1003** (2010) 018.
[10] S. Rasanen, JCAP **0902** (2009) 011.
[11] T. Buchert and J. Ehlers, Astron. Astrophys. **320** (1997) 1.
[12] A. Paranjape and T. P. Singh, JCAP **0803** (2008) 023.

[13] C. H. Chuang, J. A. Gu, and W. Y. Hwang, Class. Quant. Grav. **25** (2008) 175001.

[14] G. Lemaître, Annales Soc. Sci. Brux. Ser. ISci. Math. Astron. Phys. A **53** (1933) 51.

[15] R. C. Tolman, Proc. Nat. Acad. Sci. **20** (1934) 169.

[16] H. Bondi, Mon. Not. Roy. Astron. Soc. **107** (1947) 410.

[17] T. Biswas, R. Mansouri, and A. Notari, JCAP **0712** (2007) 017.

[18] T. Biswas and A. Notari, JCAP **0806** (2008) 021.

[19] V. F. Mukhanov, L. R. W. Abramo, and R. H. Brandenberger, Phys. Rev. Lett. **78** (1997) 1624.

[20] L. R. W. Abramo, R. H. Brandenberger, and V. F. Mukhanov, Phys. Rev. D **56** (1997) 3248.

[21] N. C. Tsamis and R. P. Woodard, Phys. Lett. B **301** (1993) 351.

[22] N. C. Tsamis and R. P. Woodard, Nucl. Phys. B **474** (1996) 235.

[23] W. Unruh, arXiv:astro-ph/9802323.

[24] H. Kodama and T. Hamazaki, Phys. Rev. D **57** (1998) 7177.

[25] G. Geshnizjani and R. Brandenberger, Phys. Rev. D **66** (2002) 123507.

[26] G. Geshnizjani and R. Brandenberger, JCAP **0504** (2005) 006.

[27] L. R. Abramo and R. P. Woodard, Phys. Rev. D **65** (2002) 043507.

[28] L. R. Abramo and R. P. Woodard, Phys. Rev. D **65** (2002) 063516.

[29] V. K. Onemli and R. P. Woodard, Class. Quant. Grav. **19** (2002) 4607.

[30] V. K. Onemli and R. P. Woodard, Phys. Rev. D **70** (2004) 107301.

[31] T. Brunier, V. K. Onemli, and R. P. Woodard, Class. Quant. Grav. **22** (2005) 59.

[32] E. O. Kahya and V. K. Onemli, Phys. Rev. D **76** (2007) 043512.

[33] E. O. Kahya, V. K. Onemli, and R. P. Woodard, Phys. Rev. D **81**, (2010) 023508.

[34] S. Deser and R. P. Woodard, Phys. Rev. Lett. **99** (2007) 111301.

[35] C. Deffayet and R. P. Woodard, JCAP **0908** (2009) 023.

[36] N. C. Tsamis and R. P. Woodard, Phys. Rev. D **81** (2010) 103509.

[37] I. L. Shapiro and J. Sola, JHEP **0202** (2002) 006; I. L. Shapiro *et al.*, Phys. Lett. B **574** (2003) 149; J. Sola and H. Stefancic, Phys. Lett. B **624** (2005) 147; I. L. Shapiro and J. Sola, Phys. Lett. B **682** (2009) 105; F. Bauer, J. Sola, and H. Stefancic, JCAP **1012** (2010) 030.

[38] A. M. Polyakov, Nucl. Phys. B **797** (2008) 199.

[39] A. M. Polyakov, Nucl. Phys. B **834** (2010) 316.

[40] D. Krotov and A. M. Polyakov, arXiv:1012.2107.

[41] N. Kumar and E. E. Flanagan, Phys. Rev. D **78** (2008) 063537.

[42] S. Weinberg, Phys. Rev. D **72** (2005) 043514.

13
Phenomenological Models

In this chapter we introduce phenomenological models for dark energy, focusing on modification for known matter components.[1]

13.1 Quintessence, Phantom and Quintom

The most well-studied parts of phenomenological models are models with rolling fields. This is because such models are direct generalizations of the cosmological constant: when the fields are not rolling, their potential energy behaves as a cosmological constant. Here we review these models in a logic order and in a brief way. The readers can find more details in the reviews [1, 2], and of course as well as the original papers.

• Quintessence.

A quintessence field [3, 4] is a scalar field with standard kinetic term, minimally coupled to gravity. The scalar field part action takes the form

$$S = \int d^4x \sqrt{-g} \left[-\frac{1}{2} g^{\mu\nu} \partial_\mu \phi \partial_\nu \phi - V(\phi) \right], \qquad (13.1)$$

where the metric convention is $(-, +, +, +)$ such that the scalar field has standard kinetic term. Take variation of $g^{\mu\nu}$, one obtains the stress tensor.

$$T_{\mu\nu} = \partial_\mu \phi \partial_\nu \phi - g_{\mu\nu} \left[\frac{1}{2} \partial^\lambda \phi \partial_\lambda \phi + V(\phi) \right]. \qquad (13.2)$$

---

[1] Sometimes it is difficult to classify whether a model belongs to modifying matter or modifying gravity because they are coupled. Indeed some models reviewed in this chapter can be thought of as modification of gravity in the infrared. But we include them here anyway because in these models gravity is not modified from the action.

The energy density and pressure can be read off from the energy-momentum tensor as

$$\rho_{\text{de}} = \frac{1}{2}\dot{\phi}^2 + V(\phi)\,, \qquad p_{\text{de}} = \frac{1}{2}\dot{\phi}^2 - V(\phi)\,. \tag{13.3}$$

As usual, the Friedmann equations are (for simplicity, let us consider the dark energy dominate case and neglect all other components)

$$3M_{\text{P}}^2 H^2 = \rho_{\text{de}}\,, \qquad -2M_{\text{P}}^2 \dot{H} = \rho_{\text{de}} + p_{\text{de}}\,, \tag{13.4}$$

where, as a reminder, $M_{\text{P}}^2 = 1/(8\pi G)$. In terms of ϕ, the above equation reads

$$3M_{\text{P}}^2 H^2 = \frac{1}{2}\dot{\phi}^2 + V\,, \qquad -2M_{\text{P}}^2 \dot{H} = \dot{\phi}^2\,. \tag{13.5}$$

Another useful equation comes out of a combination of these equations:

$$-6M_{\text{P}}^2 \left(\frac{\ddot{a}}{a}\right) = \rho_{\text{de}} + 3p_{\text{de}} = 2(\dot{\phi}^2 - V)\,. \tag{13.6}$$

From local energy conservation, the continuous equations are

$$\dot{\rho}_{\text{de}} + 3H(\rho_{\text{de}} + p_{\text{de}}) = 0\,, \qquad \ddot{\phi} + 3H\dot{\phi} + V'(\phi) = 0\,, \tag{13.7}$$

where the latter is a rewritten of the former in terms of the field ϕ.

The equation of state takes the form

$$w_{\text{de}} = \frac{p_{\text{de}}}{\rho_{\text{de}}} = \frac{\frac{1}{2}\dot{\phi}^2 - V(\phi)}{\frac{1}{2}\dot{\phi}^2 + V(\phi)}\,. \tag{13.8}$$

Note that the kinetic term has positive pressure and the potential term has negative pressure. When the field rolls slowly, the potential dominates thus w approaches -1 from above.

What kind of potentials can give rise to acceleration? A simple answer, is flat potentials. To make the answer more precise, one can have a look at the critical case: cosmic expansion without acceleration or deceleration. For this purpose, consider the power-law expansion

$$a \propto t^m\,, \tag{13.9}$$

where we have kept m general, keeping in mind that $m = 1$ corresponds to zero acceleration and $m > 1$ corresponds to acceleration. The potential driving this kind of acceleration can be solved from Eq. (13.4) as

$$V = V_0 \exp\left(-\sqrt{\frac{2}{m}}\frac{\phi}{M_\mathrm{P}}\right). \tag{13.10}$$

Thus potentials flatter than

$$V = V_0 \exp\left(-\frac{\sqrt{2}\phi}{M_\mathrm{P}}\right) \tag{13.11}$$

have acceleration solution. Moreover, as the kinetic term has much larger pressure than the potential term, the potential domination epoch is an attractor solution as long as the potential is flat.

Quintessence may be realized using axions [5] or dilatons [6], in QCD [7], in Higgs potential [8], or in unparticle theories [9].

• Phantom.

A menace from phantom [10] can be expressed in terms of the action[2]

$$S = \int d^4x \sqrt{-g}\left[\frac{1}{2}g^{\mu\nu}\partial_\mu\phi\partial_\nu\phi - V(\phi)\right]. \tag{13.12}$$

The action has a "wrong" sign kinetic term: $\mathcal{L}_\mathrm{kin} \propto -\dot{\varphi}^2$. Here phantom is also often referred to as ghost in the literature.

When expressed in terms of ρ_de and p_de, the equations of motion for phantom are identical as written in the quintessence case, while now ρ_de and p_de are expressed in terms of ϕ as

$$\rho_\mathrm{de} = -\frac{1}{2}\dot{\phi}^2 + V(\phi), \qquad p_\mathrm{de} = -\frac{1}{2}\dot{\phi}^2 - V(\phi). \tag{13.13}$$

The equation of state is

$$w_\mathrm{de} = \frac{p_\mathrm{de}}{\rho_\mathrm{de}} = \frac{\frac{1}{2}\dot{\phi}^2 + V(\phi)}{\frac{1}{2}\dot{\phi}^2 - V(\phi)}. \tag{13.14}$$

Now there are two possibilities for the equation of state: $w_\mathrm{de} > 1$ for the kinetic dominated regime and $w_\mathrm{de} < -1$ for the potential dominated regime.

---

[2] Alternatively, phantom dark energy can also be obtained in the scalar-tensor models [11].

The latter behaves as a component dark energy with super-acceleration. In other words, the universe will have an acceleration faster than exponential. The energy density keeps growing until it reaches infinity in finite proper time. When the energy density reaches infinity, the expansion rate of the universe diverges and every thing is tore off. This is called the "big rip" singularity, a physical singularity where all known physical laws break down. To see this, consider for simplicity the constant $w_{\rm de}$ case. In this case the scale factor can be written as a simple power of time as

$$a = a_0(t - t_0)^{\frac{2}{3(1+w_{\rm de})}} . \tag{13.15}$$

When $w_{\rm de} < -1$, the power of $t - t_0$ becomes negative and one concludes $a \to \infty$ when $t \to t_0$, a finite proper time for a comoving observer. The Hubble parameter $H \propto 1/(t-t_0)$ also blows up in the future. The reason is that when $w_{\rm de} < -1$, t_0 is in the future instead of in the past. Similarly the curvature blows up thus the big rip $t = t_0$ is a physical singularity. The big rip is a disaster not only for civilizations but also for physical laws, which need to be avoided or resolved.

Moreover, there is a quantum instability in the phantom models.[3] Once the phantom quanta interacts with other fields, even through gravity, there will be an instability of the vacuum because energy is no longer bounded from below. The ghost busters Cline, Jeon and Moore [13] pointed out that for the phantom to be consistent with CMB observations, the Lorentz symmetry must be broken for phantom at an energy scale lower than 3 MeV. Otherwise the effect of vacuum decay into phantom quanta and photons via gravitation could have been observed on the CMB.

• Quintom.

The quintessence field always has $w_{\rm de} > -1$ while the phantom field (as dark energy) always has $w_{\rm de} < -1$. Is it possible for a field theory model to cross the phantom divide $w_{\rm de} = -1$? The answer is positive, which is known as the quintom dark energy, a combination of the quintessence and the phantom [14].

Before reviewing what can be done, it is helpful to first have a look at what is under no-go theorems. As was noticed from the beginning [14], it is not possible to have a single scalar field to cross the phantom divide.[4] To

---

[3] However, the statement $w_{\rm de} < -1$ itself does not necessarily mean an instability. For example, holographic dark energy has $w_{\rm de} < -1$ solutions without an instability. Also in some modified gravity models such a well-behaved $w_{\rm de} < -1$ solution may be obtained [12].

[4] For counter examples on classical instabilities, see [15, 16].

see this, consider a model with time varying kinetic term $\sim f(t)\dot{\phi}^2$. If one wants to have the field cross the phantom divide when f cross 0, at $f = 0$ the field will have vanishing kinetic term thus divergent sound speed. More general proofs can be found in [17−20].

Thus one is forced to consider models with at least two fields, with the matter action

$$S = \int d^4x \sqrt{-g} \left[-\frac{1}{2}\partial^\mu \phi \partial_\mu \phi + \frac{1}{2}\partial^\mu \sigma \partial_\mu \sigma - V(\phi, \sigma) \right], \qquad (13.16)$$

The two-component dark energy has an equation of state

$$w_{\text{de}} = \frac{\frac{1}{2}\dot{\phi}^2 - \frac{1}{2}\dot{\sigma}^2 - V}{\frac{1}{2}\dot{\phi}^2 - \frac{1}{2}\dot{\sigma}^2 + V}. \qquad (13.17)$$

The model was extensively studied in [21, 2], and the references therein. Depending on the potential, the quintom equation of state may across -1 from above, or across -1 from below.[5] For an extended version of quintom scenario, see [23].

• Fast oscillating fields.

Here we consider a scalar field with standard kinetic term and potential $V \propto \phi^n$, focusing on the case that the field is oscillating around its minima and the oscillation rate is much more quickly than Hubble parameter. Turner calculated the averaged equation of state of this case [24], with

$$\langle w_{\text{de}} \rangle = \frac{n-2}{n+2}. \qquad (13.18)$$

When $n \ll 1$, the fast oscillating field can drive cosmic acceleration [25, 26]. To see this, one can apply the virial theorem. Define $A \equiv \phi\dot{\phi}$, we have

$$\dot{A} = 2T - (3H\dot{\phi} + V')\phi \simeq 2T - nV, \qquad (13.19)$$

where $T \equiv \dot{\phi}^2/2$ and we have neglected Hubble expansion in the last expression because the field is oscillating fast. Note that $\dot{A}$ is a total derivative thus the time average vanishes when we take the averaging time to be a multiple of the oscillation period. Thus

$$\langle T \rangle = \frac{n}{2} \langle V \rangle. \qquad (13.20)$$

---

[5] Alternatively, in $f(\mathcal{R})$ gravity models, w_{de} may also cross -1 [22].

Translating to the averaged equation of state, w_{de} takes the form of Eq. (13.18).

Unfortunately, the fast oscillating field has an instability, as discussed in [27]. The inhomogeneity due to the instability may get this mechanism observationally disfavored.

• Dark energy interactions.

It is possible that dark energy is dark but not lonely. Especially dark energy may decay into dark matter or vice versa. In [28–30], the following class of interaction was considered:

$$\dot{\rho}_m + 3H\rho_m = \delta\,, \tag{13.21}$$

$$\dot{\rho}_{de} + 3H(1 + w_{de})\rho_{de} = -\delta\,, \tag{13.22}$$

where δ could take the form $\delta = -b\rho_{de}$, $\delta = -b\rho_m$, or $\delta = -b(\rho_m + \rho_{de})$, etc. The effective equations of state for dark energy and matter are what one actually measures and different from their actual equations of state:

$$w_m^{eff} = -\frac{\delta}{3H\rho_m}\,, \qquad w_{de}^{eff} = w_{de} + \frac{\delta}{3H\rho_{de}}\,. \tag{13.23}$$

In some sense, the interaction could provide an explanation for the coincidence problem because if dark energy decays to dark matter in the future, the fraction of dark energy may remain at the value one observes now. However, why dark energy starts to dominate today but not much earlier, as the other half of the coincidence problem, remains unsolved.

Alternatively, the phenomenological time variation of the cosmological constant can also be implemented by a $\Lambda(t)$CDM approach, as described in [31, 32].

13.2   K-Essense, Custuton, Braiding and Ghost Condensation

Here we briefly review dark energy models from fields with modified kinetic terms.

• K-essense.

Chiba, Okabe and Yamaguchi [33] (see also Armendariz-Picon, Mukhanov and Steinhardt [34, 35]) proposed a more general framework on field theoretic dark energy, named k-essense. The idea is to generalize the kinetic term, as long as the perturbations still have second order derivative and Lorentz symmetry is preserved [36, 37]. The corresponding action in this case is

$$S = \int d^4x \sqrt{-g} \, p_{\text{de}}(X, \phi) \,, \qquad X \equiv -g^{\mu\nu} \partial_\mu \phi \partial_\nu \phi \,. \tag{13.24}$$

Here X is the conventional kinetic term. Note that p_{de} appears in the action is exactly the pressure, and the energy density takes the form

$$\rho_{\text{de}} = 2X p_{\text{de},X} - p_{\text{de}} \,, \tag{13.25}$$

where $p_{\text{de},X} \equiv \partial p_{\text{de}}/\partial X$. In terms of ρ_{de} and p_{de}, again one has the Friedmann equations (13.4).

It was shown [37] that the perturbations obey second order differential equation of motion as usual (except a special case considered later). To be ghost free, the theory should have

$$p_{\text{de},X} > 0 \,. \tag{13.26}$$

To be perturbatively stable, it is also required that the sound speed is real. The condition is

$$c_{\text{s}}^2 = \frac{p_{\text{de},X}}{p_{\text{de},X} + 2X p_{\text{de},XX}} > 0 \,. \tag{13.27}$$

Further, one might also require the sound speed $c_{\text{s}} \leqslant 1$. This is satisfied when

$$p_{\text{de},XX} \geqslant 0 \,. \tag{13.28}$$

There is a debate on whether $c_{\text{s}} \leqslant 1$ is necessary or not [38]. Thus the condition that k-essense is well behaved is Eq. (13.26) plus either Eq. (13.27) or Eq. (13.28).

To make predictions, more concrete forms of $p_{\text{de}}(X, \phi)$ are needed. The simplest case is perhaps power law k-essense, where

$$p_{\text{de}}(X, \phi) = \frac{4(1 - 3w_{\text{de}})}{9(1 + w_{\text{de}})^2 \phi^2} \left(-X + X^2\right) \,. \tag{13.29}$$

It is shown that the parameter w_{de} is indeed $w_{\text{de}} = p_{\text{de}}/\rho_{\text{de}}$, with the late time behavior $a \propto t^{\frac{2}{3(1+w_{\text{de}})}}$. Another example is DBI-essense, considered by Martin and Yamaguchi [39]. Also, k-essence may provide a unified framework of inflation and dark energy [40].

• Cuscuton.

The cuscuton dark energy was introduced by Afshordi, Chung and Geshnizjani [41], as a singular limit of k-essence. In the cuscuton model, the matter action takes the form

$$S = \int d^4x \sqrt{-g} \left[\mu^2 \sqrt{|g^{\mu\nu}\partial_\mu\phi\partial_\nu\phi|} - V(\phi) \right] . \qquad (13.30)$$

The cuscuton model has an infinite propagating speed for linear perturbations. However, the phase space volume of linear perturbations is vanishing, thus no information is propagating faster than speed of light.

Cuscuton is inspired by the plant Cuscuta, because cuscuton is parasitic in the sense that the cuscuton itself does not have its own dynamics. The equation of motion takes the form

$$(3\mu^2 H) \, \text{sgn}(\dot{\phi}) + V'(\phi) = 0 . \qquad (13.31)$$

The evolution of cuscuton follows from other energy components, which can be derived from combining Eq. (13.31) and the Friedmann equation. When adjusting $V(\phi)$, cosmic acceleration can be obtained. For example, when V is an exponential potential, the expansion history of cuscuton behaves like DGP (but perturbation theory behaves different). Cuscuton may be a minimal dynamical dark energy model because it has no dynamics, while remains the dynamical feature for dark energy. The cosmic evolution in the cuscuton model was investigated in [42].

• Kinetic gravity braiding.

K-essence is not the most general form for a scalar field with second order equation of motion. As shown in [15, 43] (see also [44, 45] in terms of a simplified yet generalized Galileon model), a more general form can be written as

$$S = \int d^4x \left[K(\phi, X) - G(\phi, X)\partial^2\phi \right] . \qquad (13.32)$$

The cosmological implications were studied in [46, 47]. It is shown that the field could behave as dark energy. This class of actions can be further generalized to [48], where cosmological implications are so far not studied.

• Ghost condensation.

As another generalization of k-essence, a model of ghost condensation is aimed to cure the quantum instability of phantom dark energy, and give an equation of state $p_{\text{de}} = -\rho_{\text{de}}$ to drive the late time acceleration of the universe. The idea is that the instabilities come from the perturbation level, where the gradient energy plays an important role. On the other hand, the spatial gradient energy does not show up in the homogeneous and isotropic background. Thus one can propose a ghost like background action, and let the spatial gradient terms cure the instability problem.

To realize this idea, Arkani-Hamed, Cheng, Luty and Mukohyama [49] (see also [50]) proposed an effective field theory of rolling ghost, preserving cosmological symmetries. The action takes the form

$$S = \int d^4x \sqrt{-g} \left[p_{\mathrm{de}}(X) + (\nabla^2\phi)^2 + \cdots \right]. \tag{13.33}$$

The cosmological solution of the model is either $\dot{\phi} = 0$ or $p_{\mathrm{de},X} = 0$. In interesting models analogous to tachyon condensation with a spontaneous symmetry breaking, the $\dot{\phi} = 0$ solution is unstable and the universe is driven to the $p_{\mathrm{de},X} = 0$ solution dynamically. This results in $X = $ constant, with $\rho_{\mathrm{de}} = -p_{\mathrm{de}}$ (recall Eq. (13.25) with $p_{\mathrm{de},X} = 0$). Thus the condensation of ghost behaves like a component of dark energy. The intuitive understanding of a constant equation of state is that, there is a shift symmetry of the system, thus rolling of ϕ with a constant speed results in a constant w. The model of ghost condensation can be thought of as an IR modification of gravity (for other IR modifications, see [51]).

Inspired by ghost condensation, an inflation model named effective field inflation [52] is proposed, with the most general action preserving cosmological symmetries. In addition, A ghost dark energy has also been proposed [53] and widely studied [54].

13.3 Higher Spin Fields

In cosmology, going to higher spin mostly means spin higher than zero. There are attempts to generalize the scalar field dark energy model to spinors [55–58], vectors [59, 60], and p-form fields [61, 62]. Here we shall briefly mention the approaches of spinors.

In curved spacetime, the action of a spinor field can be written as

$$S = \int d^4x \, e \left[\frac{i}{2} \left(\bar{\psi}\Gamma^\mu D_\mu \psi - (D_\mu \bar{\psi})\Gamma^\mu \psi \right) - V(\bar{\psi}\psi, \bar{\psi}\gamma^5\psi) \right], \tag{13.34}$$

where $e \equiv \det e^\mu{}_a$, and $e^\mu{}_a$ is the tetrad field satisfying $g_{\mu\nu}e^\mu{}_a e^\nu{}_b = \eta_{ab}$. The Γ matrices are defined as $\Gamma^\mu = e^\mu{}_a \gamma^a$. The covariant derivative is defined as $D_\mu \psi = \left(\partial_\mu + \frac{1}{2}\omega_{\mu ab}\Sigma^{ab} \right)$, with $\omega_{\mu ab} = e^\nu{}_a D_\mu e_{\nu b}$, and $\Sigma^{ab} = \frac{1}{4}[\gamma^a, \gamma^b]$. If the vacuum expectation value transforms as a scalar instead of a pseudo-scalar, the $\bar{\psi}\gamma^5\psi$ term can be dropped.

When the field is homogeneous, the energy density and pressure can be written as

$$\rho_{\rm de} = V\,, \qquad p_{\rm de} = V'\bar{\psi}\psi - V\,. \tag{13.35}$$

It can be shown from the equation of motion that $\bar{\psi}\psi \propto a^{-3}$. In other words the spinor field rolls very fast. Thus to have acceleration (and not coming simply from a cosmological constant), one need a very flat potential. For example, a potential $V \propto (\bar{\psi}\psi)^n$ with $n < 1/2$ will have acceleration. Moreover, as a spinor has multiple degrees of freedom built in, the equation of state $w_{\rm de}$ may cross -1 without divergence of sound speed [56].

For vector fields, one have to take care of constraints and large scale anisotropy. Also one has to break the conformal invariance otherwise the energy density decays too quickly. We shall not discuss these issues in details here.

13.4   Chaplygin Gas and Viscous Fluid

The matter components in cosmology are usually and conveniently written in terms of fluids. Most dark energy models have fluid description. The fluid models we refer to here correspond to the models that originate from direct modification of fluid properties.

• Chaplygin gas.

The best studied fluid model for dark energy is the Chaplygin gas model [63, 64]. The original Chaplygin gas model has the equation of state

$$p_{\rm de} = -\frac{A}{\rho_{\rm de}}\,. \tag{13.36}$$

This kind of fluid is first used in fluid mechanics which describes the air flow near the wing of an aircraft.

In the presence of matter, the energy density and pressure of the Chaplygin gas can be written as

$$\rho_{\rm de} \simeq \sqrt{A + Ba^{-6}}\,, \qquad p_{\rm de} = -\frac{A}{\sqrt{A + Ba^{-6}}}\,, \tag{13.37}$$

where B is an integration constant.

Although the Chaplygin gas itself is not motivated from field theory models, it is interesting that the model can be mimicked by a scalar field with a simple potential, in the matter or dark energy dominated universe:

$$V(\phi) = \frac{\sqrt{A}}{2} \left(\cosh 3\phi + \frac{1}{\cosh 3\phi} \right).$$ (13.38)

The Chaplygin gas model was later generalized to

$$p_{\text{de}} = -\frac{A}{\rho_{\text{de}}^{\alpha}}$$ (13.39)

in [65]. Other kinds of generalization are also possible, for example the form [66]

$$p_{\text{de}} = -\rho_{\text{de}} \left[1 - \text{sinc}(\rho_0/\rho) \right],$$ (13.40)

where $\text{sinc}(x) \equiv \sin(x)/x$, and ρ_0 is a parameter with dimension $[\text{mass}]^4$. The possible theoretical embedding and phenomenology are also considered there.

• Viscous fluid.

The local energy conservation equation (13.7) is by far satisfied by all the phenomenological models discussed above. However, it is not always the case because energy can leak to other forms such as the thermal motion from bulk viscosity. Brevik and Gorbunova noticed that the modification due to bulk viscosity leads to a model of viscous cosmology [67, 68]. The energy-momentum tensor of the fluid takes the form

$$T_{\mu\nu} = \rho u_\mu u_\nu + (p - 3H\zeta) h_{\mu\nu}, \qquad h_{\mu\nu} \equiv g_{\mu\nu} + u_\mu u_\nu.$$ (13.41)

The local energy conservation equation now reads

$$\dot{\rho} + 3H(\rho + p) = 9H^2 \zeta,$$ (13.42)

where ζ is the bulk viscosity. The equation for $\ddot{a}$ becomes

$$\frac{\ddot{a}}{a} = 12\pi G H \zeta - \frac{1 + 3w}{2} H^2.$$ (13.43)

Thus not only $w < -1/3$ could drive acceleration, a viscosity $\zeta > 0$ will also be able to drive acceleration. Meng, Ren and Hu [69, 70] pointed out that the viscous fluid may give a unified description of dark energy and dark matter, under the name "dark fluid". A more general form of such an inhomogeneous equation of state was proposed in [71].

13.5   Particle Physics Models

As is well known, non-relativistic particles behave as $w = 0$ matter and relativistic particles behave as $w = 1/3$ radiation. However, when one considers particles with exotic properties, an equation of state other than $w = 0$ or $w = 1/3$ may arise.

For example, DeDeo [72] considered a Lorentz violating action with fermion field

$$S = \int d^4x \; e \left[\bar{\psi}(i\Gamma^\mu D_\mu - m)\psi - (\bar{\psi}\gamma_5\Gamma^\mu\psi)b_\mu \right], \tag{13.44}$$

where $b_\mu = (b, 0, 0, 0)$ is a time like vector, and other notations have been clarified under Eq. (13.34). It is argued that the Lorentz violation can be achieved by coupling the spinor to a condensate of ghost.

The equation of state of the fermion particle can be calculated as

$$w_\pm(k) = \frac{k(k \mp b)}{3[m^2 + (k \mp b)^2]}, \tag{13.45}$$

where $\pm$ denotes the two helicities of the spinor, and k is the momentum of the particle. For the positive helicity particle, w becomes negative and a gas of such particles becomes a candidate for dark energy.

As another approach, Bohmer and Harko [73] reported that when a minimum length is considered, particle excitations could also arise as a component of dark energy.

Mass variation could also affect the particles' equation of state. For example, Takahashi and Tanimoto [74] showed that neutrinos with time varying mass could behave as dark energy.

Alternatively, one could modify the particles' action from symmetry considerations. For example, Stichel and Zakrzewski [75] derived an action as a dynamical realization of the zero-mass Galilean algebra with anisotropic dilational symmetry. It is shown that when the dynamical exponent is $z = 5/3$, the particles described by such an action (named darkons) behave as dark energy.

By constructing a generalized dynamics for particles, Das _et al._ [76] presented a new framework to generate an effective negative pressure and to give rise to a source for dark energy.

Another particle dark energy model is holographic gas [77], which modifies the dispersion relation of the particle such that the entropy of the gas is holographic. The model is already briefly reviewed in the chapter of Holographic principle.

In addition, Lima, Jesus and Oliveira [78] suggested that the creation of cold dark matter can yield a negative pressure and is capable to accelerate the Universe. This model can mimics the ΛCDM model, thereby can provide a good fit to current cosmological data. For more studies about this model, see [79].

13.6   Dark Energy Perturbations

A simple cosmological constant will not have perturbations simply because it is a constant. However, when the cosmological constant is replaced by other fields, fluids or particles, dark energy will typically has perturbations, either originating from the fluctuation itself or originating from the clustering of other components via gravitational interaction.

Here we follow the analysis in [80] (see also [81]), where the analysis is performed in the Newtonian gauge, or equivalently, in terms of the gauge invariant Barden potentials [82]. The perturbed metric can be written as

$$ds^2 = a(\eta)^2 \left[(1 + 2\Phi)d\eta^2 - (1 - 2\Psi)\delta_{ij}dx^i dx^j \right], \qquad (13.46)$$

where after gauge fixing ($B = 0$ and $E = 0$), Φ and Ψ are the two remaining scalar perturbation variables. Also, write the matter and dark energy components as

$$\rho \to \rho + \delta\rho \equiv \sum_i (\rho_i + \delta\rho_i), \qquad p \to p + \delta p \equiv \sum_i (p_i + \delta p_i), \qquad (13.47)$$

$$\Theta \equiv ik^i \delta T^0{}_i/(\rho + p), \qquad \Sigma \equiv \hat{k}_i \hat{k}_j \left(\delta T^i{}_j - \frac{1}{3}\delta^i{}_j \delta T^k{}_k \right) /(\rho + p). \qquad (13.48)$$

In Fourier space, the perturbed Einstein equations at linear order can be written as the following set of equations:

$$-k^2\Psi - 3\mathcal{H}(\Psi' + \mathcal{H}\Phi) = 4\pi Ga^2\delta\rho, \qquad (13.49)$$

$$k^2(\Psi' + \mathcal{H}\Phi) = 4\pi Ga^2(\rho + p)\Theta, \qquad (13.50)$$

$$\Psi'' + \mathcal{H}(2\Psi' + \Phi') + (2\mathcal{H}' + \mathcal{H}^2)\Phi + \frac{k^2}{3}(\Psi - \Phi) = 4\pi Ga^2\delta p, \qquad (13.51)$$

$$k^2(\Psi - \Phi) = 12\pi Ga^2(\rho + p)\Sigma, \qquad (13.52)$$

where prime denotes derivative with respect to the conformal time η, and $\mathcal{H} \equiv a'/a$. The above Einstein equations are consistent with a total energy conservation equation. For multiple fluids, these equations should be solved together with the equations of motion for these fluids to get a closed set of equations.

For shearless fluids (such as perfect fluid and scalar fields), the perturbation Σ vanishes and $\Psi = \Phi$. Then the Newtonian potential Φ can be solved as functions of the matter energy densities and velocities from Eqs. (13.49) and (13.50),

$$k^2\Phi = -4\pi G a^2 \left[\delta\rho + \frac{3\mathcal{H}(\rho + p)\Theta}{k^2} \right]. \qquad (13.53)$$

The solution of the above perturbation equations depends on detailed models. When w crosses the phantom divide, some equations become singular and special care is needed for proper treatments [83, 84]. In addition, dark energy perturbations also affect the cold dark matter power spectrum at large scale [85]. This can be used to distinguish dark energy models [86].

References

[1] E. J. Copeland, M. Sami, and S. Tsujikawa, Int. J. Mod. Phys. D **15** (2006) 1753.
[2] Y. F. Cai et al., Phys. Rept. **493** (2010) 1.
[3] C. Wetterich, Nucl. Phys. B **302** (1988) 668.
[4] I. Zlatev, L. M. Wang, and P. J. Steinhardt, Phys. Rev. Lett. **82** (1999) 896.
[5] S. Panda, Y. Sumitomo, and S. P. Trivedi, arXiv:1011.5877.
[6] J. P. Uzan, Phys. Rev. D **59** (1999) 123510.
[7] D. Stojkovic, G. D. Starkman, and R. Matsuo, Phys. Rev. D **77** (2008) 063006.
[8] E. Greenwood et al., Phys. Rev. D **79** (2009) 103003.
[9] D. C. Dai, S. Dutta, and D. Stojkovic, Phys. Rev. D **80** (2009) 063522.
[10] R. R. Caldwell, Phys. Lett. B **545** (2002) 23.
[11] B. Boisseau et al., Phys. Rev. Lett. **85** (2000) 2236.
[12] J. Martin, C. Schimd, and J. P. Uzan, Phys. Rev. Lett. **96** (2006) 061303.
[13] J. M. Cline, S. Jeon, and G. D. Moore, Phys. Rev. D **70** (2004) 043543.
[14] B. Feng, X. L. Wang, and X. M. Zhang, Phys. Lett. B **607** (2005) 35.
[15] C. Deffayet et al., JCAP **1010** (2010) 026.
[16] E. A. Lim, I. Sawicki, and A. Vikman, JCAP **1005** (2010) 012.
[17] A. Vikman, Phys. Rev. D **71** (2005) 023515.
[18] W. Hu, Phys. Rev. D **71** (2005) 047301.
[19] R. R. Caldwell and M. Doran, Phys. Rev. D **72** (2005) 043527.
[20] J. Q. Xia et al., Int. J. Mod. Phys. D **17** (2008) 1229.
[21] E. Elizalde, S. Nojiri, and S. D. Odintsov, Phys. Rev. D **70** (2004) 043539.
[22] K. Bamba et al., Phys. Rev. D **79** (2009) 083014; K. Bamba and C. Q. Geng, Phys. Lett. B **679** (2009) 282; K. Bamba and C. Q. Geng, Prog.

Theor. Phys. 122 (2009) 1267; K. Bamba *et al.*, Mod. Phys. Lett. A **25** (2010) 900; K. Bamba, C. Q. Geng, and C. C. Lee, arXiv:1007.0482.

[23] L. P. Chimento *et al.*, Phys. Rev. D **79** (2009) 043502.

[24] M. S. Turner, Phys. Rev. D **28** (1983) 1243.

[25] T. Damour and V. F. Mukhanov, Phys. Rev. Lett. **80** (1998) 3440.

[26] V. Sahni and L. M. Wang, Phys. Rev. D **62** (2000) 103517.

[27] M. C. Johnson and M. Kamionkowski, Phys. Rev. D **78** (2008) 063010.

[28] L. Amendola, Phys. Rev. D **62** (2000) 043511.

[29] W. Zimdahl and D. Pavon, Phys. Lett. B **521** (2001) 133.

[30] B. Wang, Y. G. Gong, and E. Abdalla, Phys. Lett. B **624** (2005) 141.

[31] J. M. Overduin and F. I. Cooperstock, Phys. Rev. D **58** (1998) 043506.

[32] P. Wang and X. H. Meng, Class. Quant. Grav. **22** (2005) 283.

[33] T. Chiba, T. Okabe, and M. Yamaguchi, Phys. Rev. D **62** (2000) 023511.

[34] C. Armendariz-Picon, V. F. Mukhanov, and P. J. Steinhardt, Phys. Rev. Lett. **85** (2000) 4438.

[35] C. Armendariz-Picon, V. F. Mukhanov, and P. J. Steinhardt, Phys. Rev. D **63** (2001) 103510.

[36] C. Armendariz-Picon, T. Damour, and V. F. Mukhanov, Phys. Lett. B **458** (1999) 209.

[37] J. Garriga and V. F. Mukhanov, Phys. Lett. B **458** (1999) 219.

[38] E. Babichev, V. Mukhanov, and A. Vikman, JHEP **0802** (2008) 101.

[39] J. Martin and M. Yamaguchi, Phys. Rev. D **77** (2008) 123508.

[40] R. Saitou and S. Nojiri, arXiv:1104.0558.

[41] N. Afshordi, D. J. H. Chung, and G. Geshnizjani, Phys. Rev. D **75** (2007) 083513.

[42] N. Afshordi *et al.*, Phys. Rev. D **75** (2007) 123509.

[43] O. Pujolas, I. Sawicki, and A. Vikman, arXiv:1103.5360.

[44] A. Nicolis, R. Rattazzi, and E. Trincherini, Phys. Rev. D **79** (2009) 064036.

[45] T. Kobayashi, M. Yamaguchi, and J. Yokoyama, Phys. Rev. Lett. **105** (2010) 231302.

[46] N. Chow and J. Khoury, Phys. Rev. D **80** (2009) 024037.

[47] S. Nesseris, A. De Felice, and S. Tsujikawa, Phys. Rev. D **82** (2010) 124054.

[48] C. Deffayet *et al.*, arXiv:1103.3260.

[49] N. Arkani-Hamed *et al.*, JHEP **0405** (2004) 074.

[50] F. Piazza and S. Tsujikawa, JCAP **0407** (2004) 004.

[51] F. Piazza, New. J. Phys. **11** (2009) 113050; S. Nesseris, F. Piazza, and S. Tsujikawa, Phys. Lett. B **689** (2010) 122.

[52] C. Cheung *et al.*, JHEP **0803** (2008) 014.

[53] F. R. Urban and A. R. Zhitnitsky, Phys. Lett. B **688** (2010) 9; F. R. Urban and A. R. Zhitnitsky, Phys. Rev. D **80** (2009) 063001; F. R. Urban and A. R. Zhitnitsky, JCAP **0909** (2009) 018; F. R. Urban and A. R. Zhitnitsky, Nucl. Phys. B **835** (2010) 135.

[54] N. Ohta, Phys. Lett. B **695** (2011) 41; R. G. Cai, arXiv:1011.3212; A. Sheykhi and M. S. Movahed, arXiv:1104.4713; A. Rozas-Fernandez, arXiv:1106.0056.

[55] M. O. Ribas, F. P. Devecchi, and G. M. Kremer, Phys. Rev. D **72** (2005) 123502.
[56] Y. F. Cai and J. Wang, Class. Quant. Grav. **25** (2008) 165014.
[57] U. A. Yajnik, arXiv:1102.2562.
[58] P. Tsyba *et al.*, arXiv:1103.5918; K. K. Yerzhanov *et al.*, arXiv:1012.3031.
[59] C. Armendariz-Picon, JCAP **0407** (2004) 007.
[60] W. Zhao and Y. Zhang, Class. Quant. Grav. **23** (2006) 3405.
[61] T. S. Koivisto and N. J. Nunes, Phys. Rev. D **80** (2009) 103509.
[62] P. Das Gupta, arXiv:0905.1621.
[63] L. R. Abramo and R. P. Woodard, Phys. Rev. D **65** (2002) 043507.
[64] A. Y. Kamenshchik, U. Moschella, and V. Pasquier, Phys. Lett. B **511** (2001) 265.
[65] M. C. Bento, O. Bertolami, and A. A. Sen, Phys. Rev. D **66** (2002) 043507.
[66] H. Hova and H. Yang, arXiv:1011.4788.
[67] I. H. Brevik, Phys. Rev. D **65** (2002) 127302.
[68] I. H. Brevik and O. Gorbunova, Gen. Rel. Grav. **37** (2005) 2039.
[69] X. H. Meng, J. Ren, and M. G. Hu, Commun. Theor. Phys. **47** (2007) 379.
[70] J. Ren and X. H. Meng, Phys. Lett. B **633** (2006) 1.
[71] S. Nojiri and S. D. Odintsov, Phys. Rev. D **72** (2005) 023003.
[72] S. DeDeo, Phys. Rev. D **73** (2006) 043520.
[73] C. G. Boehmer and T. Harko, Found. Phys. **38** (2008) 216.
[74] R. Takahashi and M. Tanimoto, Phys. Lett. B **633** (2006) 675; JHEP **0605** (2006) 021.
[75] P. C. Stichel and W. J. Zakrzewski, Phys. Rev. D **80** (2009) 083513; P. C. Stichel and W. J. Zakrzewski, Eur. Phys. J. C **70** (2010) 713.
[76] S. Das *et al.*, JHEP **0904** (2009) 115; S. Das *et al.*, arXiv:0906.1044.
[77] M. Li *et al.*, Commun. Theor. Phys. **51** (2009) 181.
[78] J. A. S. Lima, J. F. Jesus and F. A. Oliveira, JCAP **11** (2010) 027.
[79] S. Basilakos and J. A. S. Lima, Phys. Rev. D. **82** (2010) 3504; S. Basilakos M. Plionis and J. A. S. Lima, Phys. Rev. D **82** (2010) 083517; J. F. Jesus *et al.*, arXiv:1105.1027.
[80] V. F. Mukhanov, H. A. Feldman, and R. H. Brandenberger, Phys. Rept. **215** (1992) 203.
[81] C. P. Ma and E. Bertschinger, ApJ. **455** (1995) 7.
[82] J. M. Bardeen, Phys. Rev. D **22** (1980) 1882.
[83] G. B. Zhao *et al.*, Phys. Rev. D **72** (2005) 123515.
[84] J. Q. Xia *et al.*, Phys. Rev. D **73** (2006) 063521.
[85] W. Hu, Phys. Rev. D **65** (2001) 023003; J. K. Erickson *et al.*, Phys. Rev. Lett. **88** (2002) 121301.
[86] C. Gordon and W. Hu, Phys. Rev. D **70** (2004) 083003; S. Hannestad, Phys. Rev. D **71** (2005) 103519; S. Unnikrishnan, H. K. Jassal, and T. R. Seshadri, Phys. Rev. D **78** (2008) 123504; J. C. B. Sanchez and L. Perivolaropoulos, Phys. Rev. D **81** (2010) 103505; R. U. H. Ansari and S. Unnikrishnan, arXiv:1104.4609.

14
The Theoretical Challenge Revisited

Before getting to the part of experiments and fitting, it is helpful to pause and revisit the theoretical challenge. We are having too many dark energy models. On the other hand, are we close to a solution to the dark energy problem?

Let us review the questions discussed by Polchinski [1]. Firstly, Polchinski reminded us that vacuum will gravitate, at least in a local theory of gravity.

The reason is that the gravitational coupling of vacuum energy is already locally measured. This comes from the accurate measurement of equivalence principle. Aluminum and platinum have the same gravitational mass to inertial mass ratio, up to an experimental error of 10^{-12}. On the other hand, the electronic loop (see the left panel of Figure 14.1) contributions to their mass are about 10^{-3} and 3×10^{-3} to the rest energy of aluminum and platinum respectively. Thus up to an accuracy of 10^{-9}, gravity can feel the energy from the electronic vacuum loop diagram.

On the other hand, the left panel and the right panel of Figure 14.1 do not look locally different. As long as we have a local field theory, gravity cannot couple to one but not the other. Thus one can not simply ignore the gravitational coupling to vacuum energy in a serious consideration of dark energy models.

Secondly, the universe is not born empty. Instead it has a thermal history. At least we expect the abundance of helium and other light elements are explained by the big bang nucleosynthesis (BBN). Then the early universe should at least as hot as MeV in temperature. How to tune the current value of dark energy at that early time? Or if a tracking mechanism operates, how does the mechanism know the energy density will eventually fall to a small value instead of a large one?

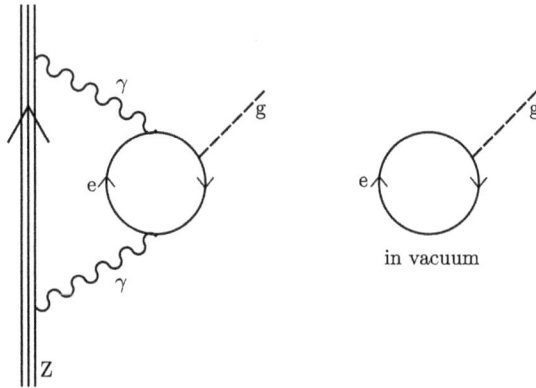

Figure 14.1   An electron self energy diagram, in the presence of an atom (left panel) and in the vacuum (right panel). In a local field theory, graviton g cannot tell the difference because it only feels the electron. It is not aware of the fields attached to the electron.

Moreover, if we expect the universe originates from a higher energy state, how does a solution of dark energy know we will eventually settle down in the present vacuum, not the $SU(2) \times U(1)$ invariant one or any other one?

Thirdly, for modifications of gravity or matter on small scales, how does this modification affect large scales while passing the solar system tests. On the other hand, for modifications of gravity or matter on larger scales, how can such a mechanism avoid the universe explode or collapse at small scales? For example, for a UV cutoff ~ 100 GeV, the curvature of the universe will be only one meter!

After these considerations, Polchinski turned to anthropic principle. However, as we have discussed in Section 7.3, the anthropic principle itself has as many problems as any of the other models, if not a lot more. Polchinski also said, "The cosmological constant is nonzero, therefore we can calculate nothing." Anthropic principle not only has trouble itself, but brings trouble to the whole field of theoretical physics.

With a great number of principles, ideas, mechanisms and models at hand, we are still not able to find a model which can answer all these questions, which at the same time is theoretically clearly derived. Thus we are either still very far away from a solution to the dark energy problem, or the solution hides behind one of the known mechanisms above, but a lot more details need to be understood. Very probably, the status of theoretical study needs to be substantially driven by future experiments, as we shall discuss below.

References

[1] J. Polchinski, arXiv:hep-th/0603249.

Part III

Observational Aspects

15
Basis of Statistics

As is known, theories and observations are connected by statistics. In this chapter, we will introduce some basic knowledge of statistics.

15.1 χ^2 Analysis

It is most common to connect theoretical models and observational data through the χ^2 statistic [1],[1] which will be briefly introduced in the following.

15.1.1 χ^2 Function

χ^2 function is a function of model parameters, and plays a key role in statistical inference. For a physical quantity ξ with experimentally measured value ξ_{obs}, standard deviation σ_ξ, and theoretically predicted value ξ_{th}, the χ^2 function is given by

$$\chi^2_\xi(\mathbf{p}) = \frac{(\xi_{\text{obs}} - \xi_{\text{th}}(\mathbf{p}))^2}{\sigma^2_\xi}, \tag{15.1}$$

where $\mathbf{p}$ denotes the model parameters. If there are many different cosmological observations giving different $\chi^2_{\xi_i}$s, the total χ^2 is the sum of all $\chi^2_{\xi_i}$s, i.e.

$$\chi^2(\mathbf{p}) = \sum_i \chi^2_{\xi_i}(\mathbf{p}). \tag{15.2}$$

It should be emphasized that Eq. (15.2) is correct only in the case where the measurements of ξ_is are independent events. In the case that the ξ_is are related to each other, the χ^2 function shall be generalized to the form

---

[1] Another alternative is the median statistic. For more details, see [2–4].

135

$$\chi^2(\mathbf{p}) = \sum_{i,j} \Delta_i (\mathrm{Cov}^{-1})_{ij} \Delta_j \,, \tag{15.3}$$

where $\Delta_i \equiv \xi_{i,\mathrm{obs}} - \xi_{i,\mathrm{th}}(\mathbf{p})$ is the data vector consisting the difference between the observational values and the theoretical values of all the ξ_is. Instead of the standard deviation σ_{ξ_i}s, a covariance matrix Cov is needed to characterize the errors of the data.

In addition to the χ^2 function, another important statistical function is the likelihood function $\mathbf{L}$ [5], which is defined as

$$\mathbf{L}(\mathbf{p}) \equiv P[\mathbf{p}|\mathbf{d}] \,, \tag{15.4}$$

where $\mathbf{d}$ denotes the observed data, and $P[x|y]$ denotes the conditional probability of x given y. The likelihood function and the χ^2 function can be related by a natural logarithm

$$\mathbf{L}(\mathbf{p}) \propto \exp\left(-\frac{1}{2}\chi^2(\mathbf{p})\right). \tag{15.5}$$

In most cases, it is more convenient to work in terms of the χ^2 function.

15.1.2 χ^2 Analysis

The χ^2 statistics has been widely used to estimate the parameters of the theoretical models on the basis of the observational data. By minimizing the χ^2 function, the best-fit model parameters can be determined as

$$\chi^2_{\min} = \chi^2(\mathbf{p}_{\mathrm{bf}}). \tag{15.6}$$

Moreover, by calculating

$$\Delta\chi^2 \equiv \chi^2 - \chi^2_{\min} \tag{15.7}$$

one can also determine the 1σ, 2σ, and 3σ (and so on) confidence level (CL) ranges of a specific model.

Statistically, for models with different n_{p} (denoting the number of free model parameters), the same CL correspond to different $\Delta\chi^2$. In general, for a model with n_{p} parameters, the value of the corresponding $\Delta\chi^2$ associated with the $n\sigma$ CL can be computed as

$$\frac{\displaystyle\int_0^{\sqrt{\Delta\chi^2}} e^{-\frac{x^2}{2}} x^{n_{\mathrm{p}}-1} dx}{\displaystyle\int_0^{+\infty} e^{-\frac{x^2}{2}} x^{n_{\mathrm{p}}-1} dx} = \text{The probability associated with the } n\sigma \text{ CL}\,,$$

$$\tag{15.8}$$

where

$$\text{The probability associated with the } n\sigma \text{ CL} = \frac{\int_0^n e^{-\frac{x^2}{2}}\,dx}{\int_0^{+\infty} e^{-\frac{x^2}{2}}\,dx}. \quad (15.9)$$

The probabilities are 68.27%, 95.45% and 99.73% for the 1σ, 2σ and 3σ CL, respectively.

In Table 15.1, we list the relationship between n_p and $\Delta\chi^2$s from $n_p = 1$ to $n_p = 10$. These $\Delta\chi^2$ values are frequently used in practical applications.

Table 15.1   Relationship between n_p and $\Delta\chi^2$

n_p	1	2	3	4	5	6	7	8	9	10
$\Delta\chi^2(1\sigma)$	1.00	2.30	3.53	4.72	5.89	7.04	8.18	9.30	10.42	11.54
$\Delta\chi^2(2\sigma)$	4.00	6.18	8.02	9.72	11.31	12.85	14.34	15.79	17.21	18.61
$\Delta\chi^2(3\sigma)$	9.00	11.83	14.16	16.25	18.21	20.06	21.85	23.57	25.26	26.90

15.2   Algorithms for the Best-Fit Analysis

Now, we will introduce how to determine the χ^2_{min} and best-fit parameters $\mathbf{p}_{bf}$. In computational science, the procedure of searching χ^2_{min} and $\mathbf{p}_{bf}$ is classified as the so-called "optimization" problem [6]. This problem has been explicitly studied, with various efficient algorithms discovered [7]. In the following, we will introduce three simple optimization algorithms that are widely used in computational science.

15.2.1   The Gradient Descent Algorithm

Gradient descent [8] is a widely used first-order optimization algorithm to seek for a *local minimum* of the given function. The idea is as follows: for a given set of parameters $\mathbf{p}_0$, the direction in which the function $\chi^2(\mathbf{p})$ decreases fastest at $\mathbf{p}_0$ is just the direction of the negative gradient. So we can make use of the formula

$$\mathbf{p}_1 = \mathbf{p}_0 - \gamma\nabla\chi^2(\mathbf{p}_0) \quad (15.10)$$

to obtain a new point $\mathbf{p}_1$. Here γ is the step size, which is a positive number. If the value of γ is small enough, we can get

$$\chi^2(\mathbf{p}_1) < \chi^2(\mathbf{p}_0). \quad (15.11)$$

To find a local minimum of the χ^2 function, one can simply generalize Eq. (15.10) to its iterative form

$$\mathbf{p}_{i+1} = \mathbf{p}_i - \gamma_i \nabla \chi^2(\mathbf{p}_i) \,, \tag{15.12}$$

so that we can get a series

$$\chi^2(\mathbf{p}_0) > \chi^2(\mathbf{p}_1) > \chi^2(\mathbf{p}_2) > \cdots . \tag{15.13}$$

Notice that the step size γ_i can vary during the process of iteration. In this way, the sequence $(\mathbf{p}_n)$ is expected to converge to the desired *local minimum*.

The efficiency of the gradient descent algorithms depends on the series γ_is. In practice, the gradient descent algorithm can be combined with a line search to determine the locally optimal step size γ on every iteration.

15.2.2 The Newton's Method

Newton's method is an algorithm widely used to find critical points of differentiable functions. Consider the Taylor expansion of the first-order derivative of the χ^2 function at a given point $\mathbf{p}$, we have

$$\nabla \chi^2(\mathbf{p} + \Delta \mathbf{p}) = \nabla \chi^2(\mathbf{p}) + \left(\mathbf{H}\chi^2(\mathbf{p})\right)\Delta \mathbf{p} \,, \tag{15.14}$$

where $\mathbf{H}\chi^2(\mathbf{p})$ is the Hessian matrix having the elements defined as

$$\left(\mathbf{H}\chi^2(\mathbf{p})\right)_{ij} = \frac{\partial^2 \chi^2(\mathbf{p})}{\partial p_i \partial p_j} \,. \tag{15.15}$$

Our goal is finding the critical points having zero first-order derivative, so we can let the L.H.S. of Eq. (15.14) be zero. Therefore, we get

$$\nabla \chi^2(\mathbf{p}) + \left(\mathbf{H}\chi^2(\mathbf{p})\right)\Delta \mathbf{p} = 0 \,, \tag{15.16}$$

as well as the iteration formula

$$\mathbf{p}_{i+1} = \mathbf{p}_i - \left(\mathbf{H}\chi^2(\mathbf{p}_i)\right)^{-1} \nabla \chi^2(\mathbf{p}_i) \,. \tag{15.17}$$

The series $\mathbf{p}_n$ is expected to converge towards a critical point of the χ^2 function.

It is also possible that finally the series converges to a maximal value of $\chi^2_{\max}$ rather than $\chi^2_{\min}$. Therefore, to make sure that we are performing an optimization, we can generalize Eq. (15.17) by including the step size γ_i,

$$\mathbf{p}_{i+1} = \mathbf{p}_i - \gamma_i \left(\mathbf{H}\chi^2(\mathbf{p}_i)\right)^{-1} \nabla \chi^2(\mathbf{p}_i) \,, \tag{15.18}$$

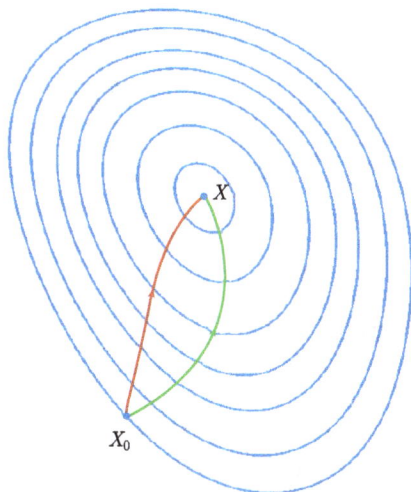

Figure 15.1   A comparison of gradient descent (green) and Newton's method (red) for minimizing a function. Compared with the gradient descent, the Newton's method can take a more direct route. See [10] for details.

and enforce a requirement $\chi^2(\mathbf{p}_{i+1}) < \chi^2(\mathbf{p}_i)$ (see [9] for more complicated Wolfe conditions).

Compared with the gradient descent method, the Newton's method uses curvature information and converges much faster towards a local minimum (see Figure 15.1). However, if the parameter space has a high dimension (n_p is large), the calculation and inversion of the Hessian matrix can be an expensive operation. One can refer to [10, 11] for some more complicated algorithms improved based on the Newton's algorithm. In addition, in the next section we will introduce the random walk algorithm as a generalization of the Newton's method, which can effectively evade the frequently calculation of the Hessian matrix and its inversion.

15.2.3   The Random Walk Algorithm

A random walk [12] is a mathematical formalization of a trajectory that consists of taking successive random steps. It can be applied to solve the best-fit searching problem.   To do this, let us start with a guessed best-fit point $\mathbf{p}_\mathrm{bf}$ and set the initial value $\chi^2_\mathrm{min} = \chi^2(\mathbf{p}_\mathrm{bf})$. Then we make a random walk around this point and get a new point

$$\mathbf{p}_\mathrm{try} = \mathbf{p}_\mathrm{bf} + \Delta\mathbf{p}\,. \tag{15.19}$$

Then, we check whether the following condition is satisfied,

$$\chi^2_{\text{try}} \equiv \chi^2(\mathbf{p}_{\text{try}}) < \chi^2_{\min}. \tag{15.20}$$

If satisfied, we can make a replacement

$$\mathbf{p}_{\text{bf}} = \mathbf{p}_{\text{try}}, \qquad \chi^2_{\min} = \chi^2_{\text{try}} \tag{15.21}$$

to get a new best-fit point. We can repeat the above operation to continuously approach the local minimum.

The efficiency of the random walk scenario depends largely on the vector $\Delta\mathbf{p}$, which represents the manner of the random walk. In fact, the random walk scenario can be combined with the gradient descent or the Newton method to obtain suitable $\Delta\mathbf{p}$s.

As an example , we will use the Newton's method and show how to generalize it to a random walk algorithm. First, let us "randomize" the $\Delta\mathbf{p}$ given by the Eq. (15.18),

$$(\mathbf{p}_{\text{try}})_i = (\mathbf{p}_{\text{bf}})_i + \gamma \times \mathbf{g}_i \times \mathbf{b}_i, \tag{15.22}$$

where the lower index i represents the i-th component of a vector. The term $\mathbf{b}_i$ denotes that the direction of the random walk is estimated from the Newton's method. It can be simply chosen as

$$\mathbf{b} = -\frac{\left(\mathbf{H}\chi^2(\mathbf{p}_{\text{bf}})\right)^{-1} \nabla\chi^2(\mathbf{p}_{\text{bf}})}{\|\left(\mathbf{H}\chi^2(\mathbf{p}_{\text{bf}})\right)^{-1} \nabla\chi^2(\mathbf{p}_{\text{bf}})\|}, \tag{15.23}$$

which is just the normalized $\Delta\mathbf{p}$ given by Eq. (15.17). The term γ is a number controlling the step size of the random walk. The vector $\mathbf{g}$ denotes the randomization of the Δp, and the value of its component can be chosen as a random number distributing around 1. In this way, the final $\Delta\mathbf{p}$ has a controllable step size and a random direction estimated from the Newton's method.

Then, we can take the following procedure:

1. Calculate $\left(\mathbf{H}\chi^2(\mathbf{p}_{\text{bf}})\right)$ at the current best-fit point $\mathbf{p}_{\text{bf}}$.
2. Repeat.
3. Use Eq. (15.22) to obtain a $\mathbf{p}_{\text{try}}$ near $\mathbf{p}_{\text{bf}}$.
4. Check the condition $\chi^2(\mathbf{p}_{\text{try}}) < \chi^2(\mathbf{p}_{\text{bf}})$. If satisfied, enforce Eq. (15.21).
5. Quit the searching until the absolute value of γ is small enough.

The number γ controls the step size of the random walk, so its value should be continually adjusted through the iteration. Compared with the Newton's method, this random walk algorithm can work much more efficient (especially n_p is large and the calculation of the χ^2 function is complicate). Similarly, it is straightforward to design a random walk algorithm based on the gradient descent method.

The random walk algorithm is a kind of Monte Carlo method [13]. As will be seen, the random walk algorithm can also be used to explore the parameter space and accomplish the whole procedures of the χ^2 analysis.

15.2.4   Summary

We have given three simple examples, the gradient descent method, the Newton's method, and the random walk method, to describe how to perform a best-fit analysis. As shown in Section 15.2.3, these algorithms can be combined together to construct a new algorithm with more efficient and reliable performance.

A limitation of these methods is that they will converge to a *local minimum* rather than the *global minimum*. If the parameter space is very complicated and has many local minimums, these algorithms usually cannot give a best result. For this case, one should explore the whole parameter space to get an overall profile. This operation is also essential, if we want to perform a complete χ^2 analysis and to estimate the errors of the model parameters. This topic will be discussed in the next section.

15.3   The Markov Chain Monte Carlo Algorithm

Next, we will discuss how to explore the whole parameter space and perform the overall χ^2 analysis.

In practice, this procedure is usually performed by using the Markov Chain Monte Carlo (MCMC) algorithms [14]. Mathematically, a chain with many elements is called Markov chain, if it is generated through a random process characterized as memoryless: the next state depends only on the current state. Moreover, MCMC method is a kind of algorithm sampling from a given probability distributions based on constructing a Markov chain. This method has an amazing feature: the samples generated through this method have the same distribution with the given probability distribution. This feature makes MCMC methods extensively used in the computational science.

In the following, first, we will introduce the "Metropolis–Hastings" algorithm [15], which is a simple MCMC algorithm based on random walk. Then we will introduce the "CosmoMC" code [16], which is the most famous MCMC code designed for the studies of cosmology.

15.3.1   The Metropolis–Hastings Algorithm

In the following, we will introduce the basic profile and procedures of the Metropolis–Hastings algorithm. let us consider a given probability distribution $P(\mathbf{p})$ with arbitrary form. Assuming that the algorithm uses a *proposal density* $Q(\mathbf{p}'; \mathbf{p}_i)$ to generate a new proposed sample $\mathbf{p}'$ depending only on the current point $\mathbf{p}$. The Metropolis–Hastings algorithm requires that, the proposed new point is accepted with probability

$$\alpha(\mathbf{p}_i, \mathbf{p}_{i+1}) = \min\left\{1, \frac{P(\mathbf{p}_{i+1}), Q(\mathbf{p}_{i+1}, \mathbf{p}_i)}{P(\mathbf{p}_i), Q(\mathbf{p}_i, \mathbf{p}_{i+1})}\right\}. \tag{15.24}$$

Thus, the transition probability (the probability of the Markov chain to move from a position in parameter space $\mathbf{p}_i$ to the next position $\mathbf{p}_{i+1}$) is

$$T_p(\mathbf{p}_i, \mathbf{p}_{i+1}) = \alpha(\mathbf{p}_i, \mathbf{p}_{i+1})Q(\mathbf{p}_i, \mathbf{p}_{i+1}). \tag{15.25}$$

This construction ensures that detailed balance holds,

$$P(\mathbf{p}_{i+1})T_p(\mathbf{p}_{i+1}, \mathbf{p}_i) = P(\mathbf{p}_i)T_p(\mathbf{p}_i, \mathbf{p}_{i+1}), \tag{15.26}$$

and hence that $P(\mathbf{p})$ is the equilibrium distribution of the Markov chain. The procedure Eq. (15.24) can be achieved by drawing a random number x from $U(0, 1)$ (the uniform distribution) and accept $\mathbf{p}_{i+1}$ if the condition

$$x < \frac{P(\mathbf{p}_{i+1}), Q(\mathbf{p}_{i+1}, \mathbf{p}_i)}{P(\mathbf{p}_i), Q(\mathbf{p}_i, \mathbf{p}_{i+1})} \tag{15.27}$$

is satisfied. Clearly, the Metropolis–Hastings algorithm is an algorithm based on random walk.

In practice, the Markov chain can be started from an arbitrary initial value $\mathbf{p}_0$, and the algorithm is run for many iterations until this initial state is "forgotten". These discarded samples are called as *burn in*. The remaining set of accepted $\mathbf{p}$s then represents a sample from the distribution $P(\mathbf{p})$ independent of the initial conditions of the Markov chain.

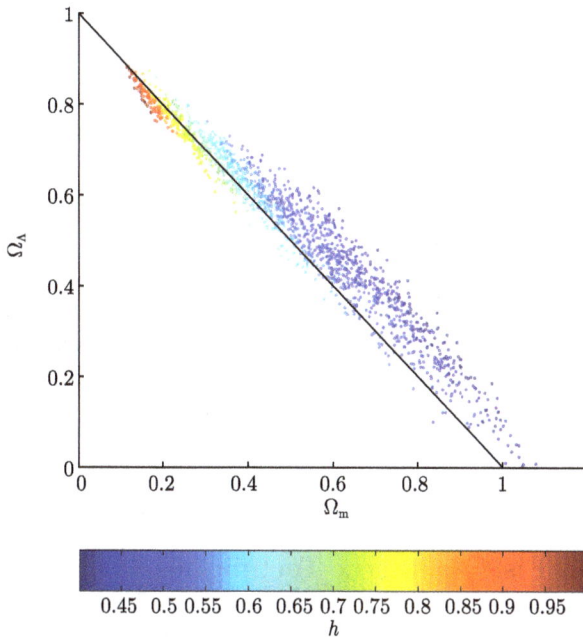

Figure 15.2   An example of 2000 samples generated by CosmoMC using the CMB data, plotted by the $\Omega_{m0}(\Omega_m$ in the figure) and $\Omega_{de0}(\Omega_\Lambda$ in the figure) values. Points are colored according to the value of h of each sample, and the solid line shows the flat universe parameters. The familiar direction of CMB degeneracy along the flat universe line is apparent. Figure from [16]. Reprinted figure with kind permission of Antony Lewis and the APS. Copyright (2002) by the American Physical Society.

In addition to the Metropolis–Hastings algorithm, there are many other MCMC algorithms widely used in computational science, e.g., the Gibbs sampling [17], the slice sampling [18], the Langevin MCMC [19], and so on. One can refer to corresponding references for details of these algorithms.

15.3.2   The CosmoMC Code

In 2002, Lewis and Bridle released the famous "CosmoMC" code [16], which provides a Fortran 90 Markov Chain Monte Carlo (MCMC) engine for exploring cosmological parameter space. The code is currently public available at [20], and is the most widely used tool to perform cosmological analysis [21].

By default, the CosmoMC code uses the Metropolis algorithm to do MCMC simulations, although other options, such as slice sampling and multicanonical sampling, are also provided. Together with the CosmoMC code for MCMC, a "getdist" program is also provided, to statistically analyze the chains obtained by CosmoMC and to output the files for the requested 1D, 2D and 3D plots. For an example of samples generated by CosmoMC, see Figure 15.2.

The CosmoMC code comprises the CAMB (Code for Anisotropies in the Microwave Background) package [22], so it is able to do the theoretical matter power spectrum and C_l calculations. The default parameters of CosmoMC include the dark energy EOS w_{de}, which is assumed constant. To study dark energy models with time-dependent EOS, one can make use of the "PPF" code [23], which supports time-dependent $w_{de}(t)$, and is publicly available at [24].

15.4   The Fisher Matrix Techniques

We have introduced the MCMC algorithm, which will take a lot of time to compute. In this section, we will introduce the Fisher Matrix formalism, which provides a quick and easy method to estimate the errors of cosmological parameters.

First, let us consider the expansion likelihood function around a fiducial model $\mathbf{p} = \mathbf{p}^*$,

$$
\ln \mathbf{L}(\mathbf{p}^* + \delta\mathbf{p}) = \ln \mathbf{L}(\mathbf{p}^*) + \sum_i \left. \frac{\partial \ln \mathbf{L}(\mathbf{p})}{\partial p_i} \right|_{\mathbf{p}=\mathbf{p}^*} \delta p_i
$$

$$
+ \frac{1}{2} \sum_{ij} \left. \frac{\partial^2 \ln \mathbf{L}(\mathbf{p})}{\partial p_i \partial p_j} \right|_{\mathbf{p}=\mathbf{p}^*} \delta p_i \delta p_j + \cdots . \quad (15.28)
$$

Clearly, the first term is just a constant. Since the fiducial model is expected (after averaging over many data realizations) to be the point of the maximum likelihood, the second term, i.e., the first derivative of the likelihood, vanishes. The third term is the Hessian of the likelihood, and is the term used in the Fisher Matrix, which is defined as the expectation value of the Hessian of the log-likelihood

$$
F_{ij} \equiv \left\langle -\frac{\partial^2 \ln \mathbf{L}}{\partial p_i \partial p_j} \right\rangle . \quad (15.29)
$$

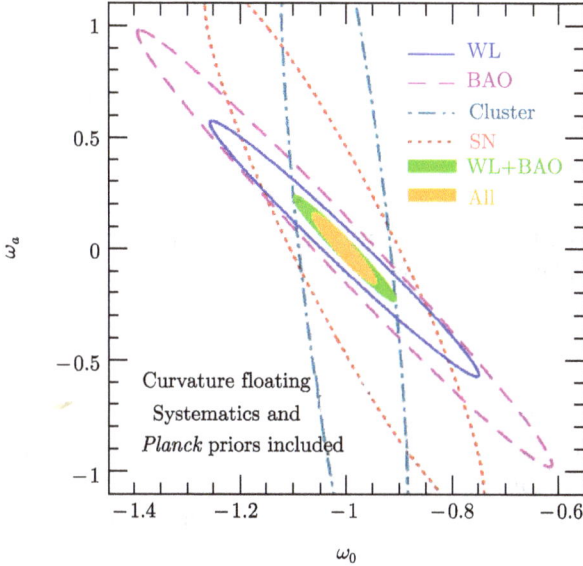

Figure 15.3   Joint w_0-w_a constraints from LSST BAO (dashed line), cluster counting (dash-dotted line), supernovae (dotted line), WL (solid line), joint BAO and WL (green shaded area), and all combined (yellow shaded area). Figure from [27]. Reprinted with kind permission of Hu Zhan and Michael Strauss.

Ignoring the 3rd or higher order terms in Eq. (15.28), then the likelihood function is fully described by the Fisher matrix plus a constant. It is clear that, in the Fisher Matrix analysis, the likelihood is assumed to have the form of multivariate Gaussian, which is just the case that models parameters are measured near the peak of the likelihood with small errors.

The Fisher Matrix is especially useful in estimating errors of parameters from given observations. Based on the Cramer–Rao inequality [25], a model parameter p_i cannot be measured to a precision better than $1/\sqrt{F_{ii}}$ when all other parameters are fixed, or a precision $\sqrt{F_{ii}^{-1}}$ when all other parameters are marginalized over (we will introduce this procedure in the following context).

The contours of constant probability are ellipsoids given by solving the equation

$$\Delta \mathbf{p}^T \tilde{F} \Delta \mathbf{p} = \Delta \chi^2 , \qquad (15.30)$$

where $\Delta \mathbf{p} = \mathbf{p} - \mathbf{p}^*$ is the parameter vector around the fiducial model $\mathbf{p}^*$, and $\Delta \chi^2$ is a constant determined by Eq. (15.9). The matrix $\tilde{F}$ is the "marginalized" Fisher matrix of the parameters of interest (all the parameters that are not of interest are marginalized over). The calculation is very simple: invert F, remove the rows and columns that are being marginalized over, and then invert the result to obtain the marginalized Fisher matrix [26].

It is useful to define a value of the so-called "figure of merit", which is proportional to the area of the error ellipse defined by Eq. (15.30). For example, the DETF figure of merit is defined as the reciprocal of the area of the error ellipse enclosing the 95% CL in the w_0-w_a (the two parameters in the dark energy EOS w) plane. Larger figure of merit indicates greater accuracy of the considered observations.

Although the Fisher Matrix formalism is not as powerful and accurate as the MCMC algorithm, it is much easier to calculate in practice. So it is widely used to forecast errors and design experiments (see Figure 15.3 for an example). For more details about the Fisher Matrix, one can refer to [28, 29].

References

[1] S. Dodelson, *Modern Cosmology* (Academic Press, 2003).
[2] J. Richard Gott III *et al.*, ApJ. **549** (2001) 1.
[3] P. P. Avelino, C. J. A. P. Martins, and P. Pinto, ApJ. **575** (2002) 989.
[4] A. Barreira and P. P. Avelino, arXiv:1107.3971.
[5] A. W. F. Edwards, *Likelihood* (Cambridge University Press, Cambridge, 1972); Yudi Pawitan, in *All Likelihood: Statistical Modelling and Inference Using Likelihood* (Oxford University Press, 2001).
[6] S. Bradley, A. Hax, and T. Magnanti, *Applied Mathematical Programming* (Addison Wesley, 1977); R. L. Rardin, *Optimization in Operations Research* (Prentice Hall, 1997).
[7] http://en.wikipedia.org/wiki/Category_Optimization_algorithms.
[8] J. A. Snyman, *Practical Mathematical Optimization: An Introduction to Basic Optimization Theory and Classical and New Gradient-Based Algorithms* (Springer Publishing, 2005).
[9] http://en.wikipedia.org/wiki/Wolfe_conditions.
[10] http://en.wikipedia.org/wiki/Newton%27s_method_in_optimization.
[11] http://en.wikipedia.org/wiki/Quasi-Newton_method.
[12] K. Pearson, Nature **72** (1905) 294.
[13] H. L. Anderson, *Metropolis, Monte Carlo and the MANIAC*, Los Alamos Science **14** (1986) 96.
[14] D. Gamerman, *Markov Chain Monte Carlo: Stochastic Simulation for Bayesian Inference* (Chapman and Hall, 1997); Bernd A. Berg, *Markov*

Chain Monte Carlo Simulations and Their Statistical Analysis (World Scientific, Hackensack, NJ, 2004).

[15] N. Metropolis, A. W. Rosenbluth, M. N. Rosenbluth, A. H. Teller, and E. Teller, Journal of Chemical Physics **21** (1953) 1087.

[16] A. Lewis and S. Bridle, Phys. Rev. D **66** (2002) 103511.

[17] S. Geman and D. Geman, IEEE Transactions on Pattern Analysis and Machine Intelligence **6** (1984) 721.

[18] R. M. Neal, Annals of Statistics **31** (2003) 705.

[19] O. Stramer and R. Tweedie, Methodology and Computing in Applied Probability **1** (1999) 307.

[20] http://cosmologist.info/cosmomc/.

[21] D. Rapetti *et al.*, MNRAS **360** (2005) 555; P. McDonald *et al.*, ApJ. **635** (2005) 761; P. McDonald *et al.*, Astrophys. J. Suppl. **163** (2006) 80; L. Perotto *et al.*, JCAP **0610** (2006) 013; J. Lesgourgues, M. Viel, M. G. Haehnelt and R. Massey, JCAP **0711** (2007) 008; S. W. Allen *et al.*, MNRAS **383** (2008) 879.

[22] A. Lewis, A. Challinor, and A. Lasenby, ApJ. **538** (2000) 473.

[23] W. Hu and I. Sawicki, Phys. Rev. D **76** (2007) 104043; W. Hu, Phys. Rev. D **77** (2008) 103524; W. Fang *et al.*, Phys. Rev. D **78** (2008) 103509; W. Fang, W. Hu and A. Lewis, Phys. Rev. D **78** (2008) 087303.

[24] http://background.uchicago.edu/ppf/.

[25] C. R. Rao, Bulletin of the Calcutta Mathematical Society **37** (1945) 81; H. Cramer, *Mathematical Methods of Statistics* (Princeton University Press, Princeton, NJ, 1946).

[26] A. Albrecht *et al.*, arXiv:astro-ph/0609591.

[27] P. A. Abell *et al.*, arXiv:0912.0201.

[28] G. Jungman *et al.*, Phys. Rev. D **54** (1996) 1332.

[29] B. A. Bassett *et al.*, arXiv:0906.0993.

16
Cosmic Probes of Dark Energy

The most common approach to probe dark energy is through its effect on the expansion history of the universe. This effect can be detected via the luminosity distance $d_L(z)$ and the angular diameter distance $D_A(z)$. In addition, the growth of large-scale structure can also provide useful constraints on dark energy. In this chapter, we will review some mainstream cosmological observations, introduce the basic principles of these observations, and describe how they are introduced into the χ^2 statistics.

16.1 Type Ia Supernovae

Type Ia supernovae (SNIa) is a sub-category of cataclysmic variable stars that results from the violent explosion of a white dwarf star. A white dwarf star can accrete mass from its companion star; as it approaches the Chandrasekhar mass, the thermonuclear explosion will occur [1]. Therefore, SNIa can be used as standard candles to measure the luminosity distance $d_L(z)$ [2–4], and thus provides a useful tool to measure the expansion history of the universe. In 1998, using 16 distant and 34 nearby SNIa from the Hubble space telescope (HST) observations, Riess *et al.* [5] first discovered the acceleration of expanding universe. Soon after, based on the analysis of 18 nearby supernova (SN) from the Calan–Tololo sample and 42 high-redshift SN, Perlmutter *et al.* confirmed the discovery of cosmic acceleration [6]. The discovery of the universe's accelerating expansion (see Figure 16.1) was another big surprise since Edwin Hubble discovered the cosmic expansion in 1929. Because of the great discovery of cosmic acceleration, Saul Perlmutter, Brian Schmidt, and Adam Riess won the Nobel prize in physics 2011. For a more detailed history about the discovery of cosmic acceleration, see [8].

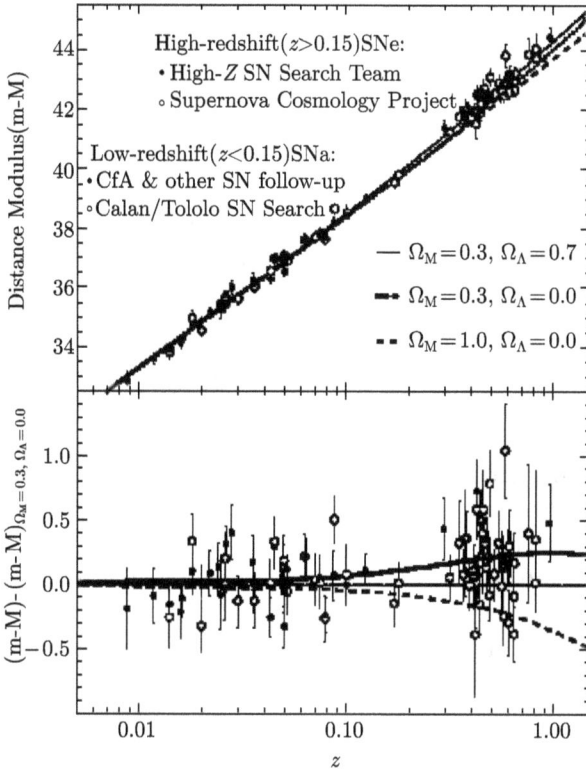

Figure 16.1   Discovery data: Hubble diagram of SNIa measured by the Supernova Cosmology Project and the High-Z Supernova Team. Lower panel shows residuals in distance modulus relative to an open universe with $\Omega_{m0}(\Omega_M) = 0.3$ and $\Omega_{de0}(\Omega_\Lambda) = 0$. From [7], based on [5, 6]. Reprinted with kind permission of Dragan Huterer and the Annual Reviews.

In recent years, the surveys of SNIa have drawn more and more attention [7, 9, 10]. Many research groups focused on this field, such as the Higher-Z Team [11, 12], the Supernova Legacy Survey (SNLS) [13, 14], the ESSENCE (denoting "Equation of State: SupErNovae trace Cosmic Expansion") [15], the Nearby Supernova Factory (NSF) [16], the Carnegie Supernova Project (CSP) [17], the Lick Observatory Supernova Search (LOSS) [18], and the Sloan Digital Sky Survey (SDSS) SN Survey [19], etc. In 2008, the Supernova Cosmology Project (SCP) provided a framework to analyze these SNIa datasets in a homogeneous manner. From 414 SNIa samples, they selected 307 high-quality SNIa to compose the "Union" dataset [20].

In 2009, the Center for Astrophysics (CfA) SN project combined 90 low-redshift SNIa samples with the Union dataset and obtained the "Constitution" sample [21]. In 2010, the SCP updated their SNIa sample, and released the "Union2" dataset [22], which consisting of 557 SNIa. Moreover, in a latest work [23], the Union2.1 SNIa dataset, which consisting of 580 SNIa, was released.

To constrain dark energy by using the SNIa data, the absolute magnitude of SNIa must be determined first. Since the detailed mechanism of SNIa explosions remains uncertain [24], SNIa are not intrinsically standard candles, with a 1σ spread of order 0.3 mag in peak B band luminosity [25]. Fortunately, in 1992 Phillips found that SNIa has a clear correlation between their intrinsic brightness at maximum light and the duration of their light curve [26]. This so-called "Phillips correlation" was then used to turn SNIa into standard candles [27]. In 2004, utilizing "stretch" duration-magnitude correction, Kim *et al.* [28] reduced the dispersion on SNIa peak magnitude into only 0.10–0.15 mag. Figure 16.2 shows a comparison of SNIa light curves before and after applying the "stretch" duration-magnitude correction.

After determining the absolute magnitude of the SNIa, one can obtain the observational distance modulus

$$\mu_{\text{obs}} = m - M \,, \tag{16.1}$$

where m is the apparent magnitude, and M is the absolute magnitude. On the other hand, the theoretical distance modulus can be calculated as

$$\mu_{\text{th}}(z_i) \equiv 5 \log_{10} d_{\text{L}}(z_i) + 25 \,, \tag{16.2}$$

and the luminosity distance $d_{\text{L}}(z)$ is

$$d_{\text{L}}(z) = \frac{1+z}{H_0 \sqrt{|\Omega_{k0}|}} f_k \left(H_0 \sqrt{|\Omega_{k0}|} \int_0^z \frac{dz'}{H(z')} \right) \,, \tag{16.3}$$

where

$$f_k(x) = \begin{cases} \sin(x), & \text{if } \Omega_{k0} < 0 \ (k=1); \\ x, & \text{if } \Omega_{k0} = 0 \ (k=0); \\ \sinh(x), & \text{if } \Omega_{k0} > 0 \ (k=-1) \,. \end{cases} \tag{16.4}$$

The χ^2 for the SNIa data is

$$\chi_{\text{SN}}^2(\mathbf{p}) = \sum_{i=1}^{n} \frac{[\mu_{\text{obs}}(z_i) - \mu_{\text{th}}(z_i; \mathbf{p})]^2}{\sigma_i^2} \,, \tag{16.5}$$

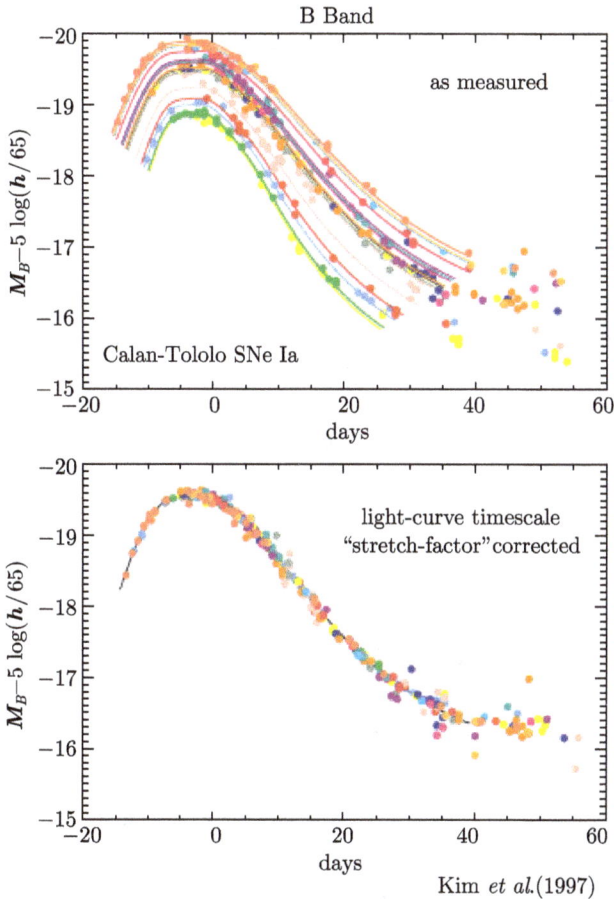

Figure 16.2   *Upper panel*: B band light curves of the Calan–Tololo SNIa sample before any duration-magnitude correction. *Lower panel*: Same light curves after applying the "stretch" duration-magnitude correction of [28]. Reprinted with kind permission of Alex G. Kim.

where **p** denotes the model parameters, $\mu_{obs}(z_i)$ and σ_i are the observed value and the corresponding 1σ error of distance modulus for each SNIa, respectively. It should be mentioned that due to the uncertainty in the absolute magnitude of a SNIa, the degeneracy between the Hubble constant and the absolute magnitude implies that one cannot quote constraints on either one. Thus, when dealing with SNIa data, people often analytically marginalize the nuisance parameter H_0 [29].

It should be stressed that the Eq. (16.5) only includes the statistical errors of SNIa, while the systematic errors of SNIa are ignored. The systematic errors come from various factors including the errors in the photometry, the calibration, the identification of SNIa, the selection bias, the intrinsic variation of physical properties of SNIa, the host-galaxy extinction, the gravitational lensing, and so on. Currently, the systematic errors in the SNIa data have been comparable with the statistical errors, thus they should be considered seriously. To include the effect of systematic errors in the analysis, a prescription for using the Union2 compilation has been provided in [30]. The key of this prescription is a 557×557 systematics covariance matrix, $\mathrm{Cov_{Union2}}$, which captures the systematic errors from SNIa (This covariance matrix with systematics can be downloaded from [30]). Utilizing $\mathrm{Cov_{Union2}}$, one can calculate the following quantities

$$A(\mathbf{p}) = (\mu_i^{\mathrm{obs}} - \mu_i^{\mathrm{th}}(\mu_0 = 0, \mathbf{p}))(\mathrm{Cov_{Union2}^{-1}})_{ij}(\mu_j^{\mathrm{obs}} - \mu_j^{\mathrm{th}}(\mu_0 = 0, \mathbf{p})),$$

(16.6)

$$B(\mathbf{p}) = \sum_{i=1}^{557} (\mathrm{Cov_{Union2}^{-1}})_{ij}(\mu_j^{\mathrm{obs}} - \mu_j^{\mathrm{th}}(\mu_0 = 0, \mathbf{p})),$$

(16.7)

$$C = \sum_{i,j=1}^{557} (\mathrm{Cov_{Union2}^{-1}})_{ij},$$

(16.8)

where $\mu_i^{\mathrm{obs}} = \mu_{\mathrm{obs}}(z_i)$, $\mu_i^{\mathrm{th}} = \mu_{\mathrm{th}}(z_i)$. The χ^2 for the SNIa data is given by

$$\chi^2_{\mathrm{SN}}(\mathbf{p}) = A(\mathbf{p}) - \frac{B(\mathbf{p})^2}{C}.$$

(16.9)

In a latest work [31], by using the SNLS3 sample of 472 SN, Conley *et al.* gave a most careful and detailed study about the systematic uncertainties of SNIa. After taking into account the systematic uncertainties of SNIa, the χ^2 function can be written as

$$\chi^2_{\mathrm{SN}} = \Delta \vec{m}^T \cdot \mathrm{Cov_{SNLS3}^{-1}} \cdot \Delta \vec{m},$$

(16.10)

where $\mathrm{Cov_{SNLS3}}$ is a 472×472 covariance matrix capturing the statistic and systematic uncertainties of the SNIa sample, $\Delta \vec{m} = \vec{m}_{\mathrm{B}} - \vec{m}_{\mathrm{mod}}$ is a vector of model residuals, $\vec{m}_{\mathrm{B}}$ is the rest-frame peak B band magnitude of an SN, and $\vec{m}_{\mathrm{mod}}$ is the predicted magnitude of the SN given by the theoretical model and two other quantities (stretch and color) describing the light-curve of the particular SN. The model magnitude m_{mod} is given by

$$m_{\mathrm{mod}} = 5 \log_{10} D_{\mathrm{L}}(z_{\mathrm{hel}}, z_{\mathrm{cmb}}) - \alpha(s-1) + \beta \mathcal{C} + \mathcal{M}, \qquad (16.11)$$

where D_{L} is the Hubble-constant free luminosity distance, z_{cmb} and z_{hel} are the CMB frame and heliocentric redshifts of the SN, s is the stretch measure for the SN, and $\mathcal{C}$ is the color measure for the SN. α and β are nuisance parameters which characterize the stretch-luminosity and color-luminosity relationships, respectively. $\mathcal{M}$ is another nuisance parameter representing some combination of the absolute magnitude of a fiducial SNIa and the Hubble constant. The total covariance matrix $\mathrm{Cov}_{\mathrm{SNLS3}}$ in Eq. (16.10) captures both the statistical and systematic uncertainties of the SNIa data. One can factor it as [31],

$$\mathrm{Cov}_{\mathrm{SNLS3}} = D_{\mathrm{stat}} + \mathrm{Cov}_{\mathrm{stat}} + \mathrm{Cov}_{\mathrm{sys}}, \qquad (16.12)$$

where D_{stat} is the purely diagonal part of the statistical uncertainties, $\mathrm{Cov}_{\mathrm{stat}}$ is the off-diagonal part of the statistical uncertainties, and $\mathrm{Cov}_{\mathrm{sys}}$ is the part capturing the systematic uncertainties. It should be mentioned that, for different α and β, these covariance matrices are also different. Therefore, one has to reconstruct the covariance matrix $\mathrm{Cov}_{\mathrm{SNLS3}}$ for the corresponding values of α and β. Here we do not describe these covariance matrices one by one. One can refer to [31, 32] and the public code [33] for more details about the explicit forms of these covariance matrices and the calculation of χ^2_{SN}.

In addition, some interesting methods are also proposed to reduce the systematic errors in the SNIa data. For instance, Wang and colleagues [34] developed a consistent frameworks for the flux-averaging of SNIa to reduce the effect of weak lensing on SNIa data. Currently, the systematic errors in the SNIa observations are the major factors that confining their ability to precisely measure the properties of dark energy. To enhance the precision of SNIa data, improvements on the photometric technique, as well as better understandings of the dust absorption and the SN explosions, are needed. For more details on SNIa observation and its cosmological applications, see [35, 36] and references therein.

16.2   Cosmic Microwave Background

CMB is the legacy of the cosmic recombination epoch. It contains abundant information of the early universe. In 1964 CMB was firstly detected by Penzias and Wilson [37], who received the Nobel prize in physics in 1978. Their work provided strong evidence that supports the Big Bang theory of

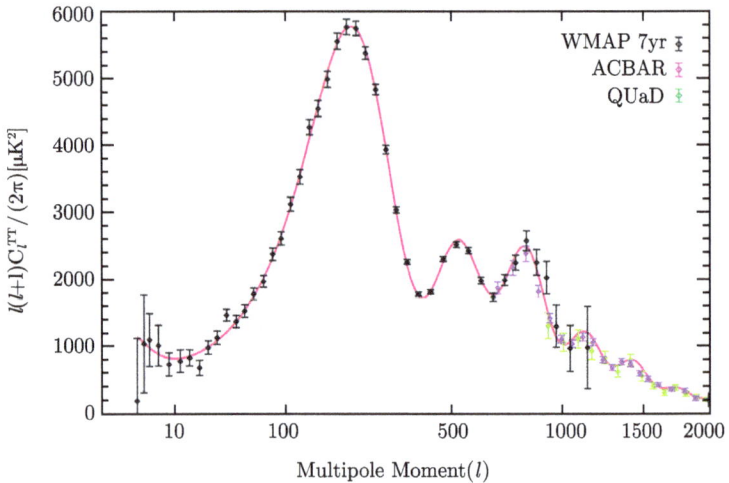

Figure 16.3   The WMAP 7-year temperature power spectrum, along with the temperature power spectra from the ACBAR [50] and QUaD [51] experiments. The solid line shows the best-fitting ΛCDM model to the WMAP data, corresponding to $\Omega_{de0} = 0.738$. From [52]. Reproduced by permission of Eiichiro Komatsu and the AAS.

the universe [38]. In 1989, the first generation of CMB satellite, the Cosmic Background Explorer (COBE), was launched. It discovered the CMB anisotropy for the first time [39] and opened the era of the precise cosmology. Two of COBE's principal investigators, Smoot and Mather, received the Nobel prize in physics in 2006. In 1999, the TOCO, BOOMERang, and Maxima experiments [40] firstly measured the acoustic oscillations in the CMB anisotropy angular power spectrum [41–47] In 2001, the second generation of CMB satellite, the Wilkinson Microwave Anisotropy Probe (WMAP) [48, 49], was launched. It precisely measured the CMB spectrum (for latest results of WMAP, see Figure 16.3) and probed various cosmological parameters with a higher accuracy. Recently, the Planck satellite, as the successor to WMAP, was launched in 2009. The early results of Planck have been released recently [53].

The positions of the CMB acoustic peaks lie on the expand history from the decoupling epoch to the present epoch, and contains the information of dark energy thereof. Two distance ratios are often used to constrain dark energy. The first distance ratio is the so-called "acoustic scale" l_A, which represents the CMB multipole corresponds to the location of the acoustic peak. It can be calculated as

$$l_A = (1 + z_*) \frac{\pi D_A(z_*)}{r_s(z_*)}. \tag{16.13}$$

Here z_* denotes the redshift of the photon decoupling epoch, whose fitting formula is given by [46]

$$z_* = 1048[1 + 0.00124(\Omega_{b0}h^2)^{-0.738}] \left[1 + g_1(\Omega_{m0}h^2)^{g_2}\right], \tag{16.14}$$

where

$$g_1 = \frac{0.0783(\Omega_{b0}h^2)^{-0.238}}{1 + 39.5(\Omega_{b0}h^2)^{0.763}}, \qquad g_2 = \frac{0.560}{1 + 21.1(\Omega_{b0}h^2)^{1.81}}. \tag{16.15}$$

$D_A(z)$ is the proper angular diameter distance,

$$D_A(z) = \frac{1}{H_0} \frac{f_k\left[H_0\sqrt{|\Omega_{k0}|} \int_0^z \frac{dz'}{H(z')}\right]}{(1 + z)\sqrt{|\Omega_{k0}|}}, \tag{16.16}$$

and r_s is the comoving sound horizon size,

$$r_s(z) = \int_0^{t_{rec}} c_s(1+z)dt = \frac{1}{\sqrt{3}} \int_0^{1/(1+z)} \frac{da}{a^2 H(a)\sqrt{1 + (3\Omega_{b0}/4\Omega_{\gamma 0})a}}. \tag{16.17}$$

One can calculate $\chi^2_{CMB} = (l_A^{obs} - l_A^{th})^2/\sigma_{l_A}^2$ to include the CMB data into the χ^2 statistics. The second distance ratio is the so-called "shift parameter" R which takes the form (16.18)

$$R(z_*) = \sqrt{\Omega_{m0}}H_0(1 + z_*)D_A(z_*). \tag{16.18}$$

One can also calculate $\chi^2_{CMB} = (R_{obs} - R_{th})^2/\sigma_R^2$ to reflect the contribution of the CMB data.

Here we list the values of $l_A(z_*)$, $R(z_*)$ and z_* obtained from the WMAP 7-year observations (Table 9 of [52]) in Table 16.1. One can use these data to calculate the χ^2_{CMB} defined above. Moreover, the χ^2_{CMB} can be calculated as

$$\chi^2_{CMB} = (x_i^{obs} - x_i^{th})(Cov_{CMB}^{-1})_{ij}(x_j^{obs} - x_j^{th}), \tag{16.19}$$

Table 16.1   Distance Priors from WMAP 7-year Fit

	7-year ML[a]	7-year Mean[b]	Error, σ
$l_A(z_*)$	302.09	302.69	0.76
$R(z_*)$	1.725	1.726	0.018
z_*	1091.3	1091.36	0.91

[a]Maximum likelihood values; [b]Mean of the likelihood.

where $x_i = (l_A, R, z_*)$ is a vector, and (Cov_{CMB}^{-1}) is the inverse covariance matrix. For the WMAP 7-year observations [52], the inverse covariance matrix takes the following forms

$$\text{Cov}_{CMB}^{-1} = \begin{pmatrix} 2.305 & 29.698 & -1.333 \\ & 6825.270 & -113.180 \\ & & 3.414 \end{pmatrix}.$$

In addition, the presence of dark energy also affects the large scale anisotropy of the CMB through the integrated Sachs–Wolfe (ISW) effect [54], which is not captured by Eq. (16.19). The ISW effect is caused by the variation of the gravitational potential during the epoch of the cosmic acceleration. It provides independent evidence for the existence of dark energy [55, 56]. The ISW effect can be detected through the cross correlation between the CMB and LSS [57]. In [58], Ho $et\ al.$ reported a 3.7σ detection of ISW by cross-correlating the SDSS LSS observations with the WMAP CMB anisotropies results. For more details on CMB observation and its cosmological applications, see [59–61] and references therein.

16.3   Baryon Acoustic Oscillations

Baryon acoustic oscillation (BAO) refers to an overdensity or clustering of baryonic matter at certain length scales due to acoustic waves which propagated in the early universe [41, 62, 63]. Similar to SNIa, which provides a "standard candle" for astronomical observations, the BAO, which happens on some typical scales, provides a "standard ruler" for length scale in cosmology to explore the expansion history of the universe [64]. The length of this standard ruler (~ 150 Mpc in today's universe [65]) corresponds to the distance that a sound wave propagating from a point source at the end of inflation would have traveled before decoupling. BAO has a characteristic imprint on the matter power spectrum [42, 43]. So it can be measured at low redshifts $z < 1$ through the astronomical surveys of galaxy clusters [66].

In addition, BAO scales can also be measured through 21 cm emission from reionization, which provides abundant information about the early universe at high redshifts $1.5 \leqslant z \leqslant 20$ [67]. The apparent size of the BAO measured from astronomical observations then leads to the measurements of the Hubble parameter $H(z)$ and the angular diameter distance $D_A(z)$ [68].

The BAO measurement does not require precision measurements of galaxy magnitudes, nor does it require that galaxy images be resolved; instead, only their three-dimensional positions need to be determined. So the BAO observations are less affected by astronomical uncertainties than other probes of dark energy, although it still suffers from some systematic uncertainties, such as the effects of non-linear gravitational evolution and redshift distortions of clustering [69, 70]. The acoustic signature of BAO has already been obtained in the galaxy power spectrum at low redshift [71, 64]. For examples, the Two-degree-Field Galaxy Redshift Survey (2dFGRS) [72], which was operated by the Anglo-Australian Observatory (AAO) between 1997 and 2002, had made public their BAO data in 2003. In addition, using a dedicated 2.5-m wide-angle optical telescope, the Sloan Digital Sky Survey (SDSS) [73], was launched in 2000. In 2006, the survey entered a new phase, the SDSS-II survey. In 3 years, it completed the observations of a huge contiguous region of the northern skies. The BAO power spectrum measured by the SDSS-II survey [74] is shown in Figure 16.4. In 2009, the SDSS-III survey begun. In January 2011, SDSS-III publicly released its eighth Data Release (DR8) [75], which is the latest BAO dataset to data.

From measurements of the galaxy clusters, the BAO scales in both transverse and line-of-sight directions are obtained; they correspond to the quantities $r(z)/r_s(z)$ and $r_s(z)/H(z)$, respectively. In addition, three characteristic quantities of BAO, including the A parameter, $D_V(0.35)/D_V(0.2)$, and $r_s(z_d)/D_V(z)$, are often used to constrain dark energy parameters. In the following we will provide a rough introduction to these three quantities.

The A parameter is defined as [64]

$$A_{\text{th}} = \frac{\sqrt{\Omega_{m0}}}{(H(z_b)/H_0)^{\frac{1}{3}}} \left[\frac{1}{z_b\sqrt{|\Omega_{k0}|}} f_k\left(H_0\sqrt{|\Omega_{k0}|} \int_0^{z_b} \frac{dz'}{H(z')} \right) \right]^{\frac{2}{3}}, \quad (16.20)$$

where z_b is the corresponding redshift of the BAO experiment, for SDSS luminous red galaxies (LRG), $z_b=0.35$. The SDSS BAO measurements [64] gave $A_{\text{obs}} = 0.469(n_s/0.98)^{-0.35} \pm 0.017$, where the WMAP 7-year

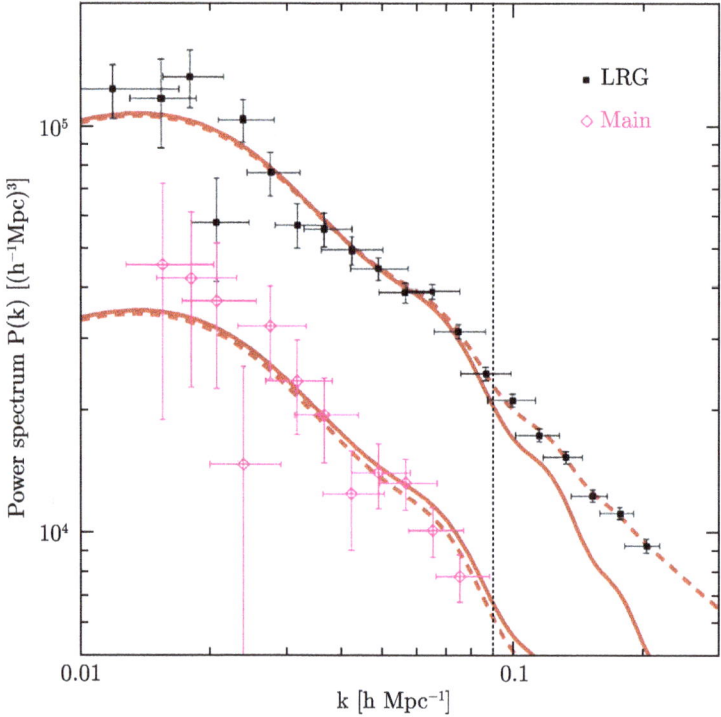

Figure 16.4   The BAO power spectrum for the full luminous red galaxy (LRG) and main galaxy samples, measured by the SDSS-II survey. The solid curves correspond to the linear theory ΛCDM fits to WMAP3. The dashed curves include the nonlinear corrections. Results from LRG and main galaxy samples are consistent with each other, and all provide a confirmation of the predicted large-scale ΛCDM power spectrum. When combined with the WMAP3 result, the LRG data yield to $\Omega_{\mathrm{de}0} = 0.761^{+0.017}_{-0.018}$ and $w_{\mathrm{de}} = -0.941^{+0.087}_{-0.101}$. From [74]. Reprinted figure with kind permission of Max Tegmark and the APS. Copyright (2006) by the American Physical Society.

results gave a best-fit value $n_{\mathrm{s}} = 0.963$. One can calculate $\chi^2_{\mathrm{BAO}} = (A_{\mathrm{obs}} - A_{\mathrm{th}})^2/\sigma^2_A$ to reflect the impact of the BAO data. The A parameter is considered independent of dark energy models, and has been widely used in the literature.

Next, let us turn to the quantity $D_V(0.35)/D_V(0.2)$. Since the current observations are not sufficient enough to measure the BAO scales in both transverse and line-of-sight directions independently, alternatively people construct an effective distance ratio $D_V(z)$, which is defined as [64]

$$D_V(z) = \left[(1+z)^2 D_A^2(z) \frac{z}{H(z)}\right]^{1/3}. \tag{16.21}$$

In 2005, Eisenstein *et al.* [64] provided a constraint $D_V(z) = 1370 \pm 64$ Mpc at the redshift $z=0.35$. In 2009, the SDSS DR7 sample [76] gave $D_V(0.35)/D_V(0.2) = 1.736 \pm 0.065$. One can use this quantity to reflect the impact of the BAO data.

Lastly, we introduce the quantity $r_s(z_d)/D_V(z)$. From the SDSS and the 2dFGRS, one can extract a quantity $r_s(z_d)/D_V(z)$ at given z, where z_d denotes the redshift of the drag epoch, whose fitting formula was proposed by Eisenstein and Hu [45],

$$z_d = \frac{1291(\Omega_{m0}h^2)^{0.251}}{1+0.659(\Omega_{m0}h^2)^{0.828}}\left[1 + b_1(\Omega_{b0}h^2)^{b_2}\right], \tag{16.22}$$

where

$$b_1 = 0.313(\Omega_{m0}h^2)^{-0.419}\left[1+0.607(\Omega_{m0}h^2)^{0.674}\right],$$
$$b_2 = 0.238(\Omega_{m0}h^2)^{0.223}. \tag{16.23}$$

It is widely believed that $r_s(z_d)/D_V(z)$ contains more information of BAO than the previous two quantities.

In addition, one can also use the covariance matrix method to construct the χ^2 function of the BAO data

$$\chi^2_{BAO} = \Delta p_i[\mathrm{Cov}_{BAO}^{-1}(p_i, p_j)]\Delta p_j, \qquad \Delta p_i = p_i^{\mathrm{data}} - p_i. \tag{16.24}$$

For the latest BAO data from SDSS DR7 [76], $p_1 = d_{0.2}$ and $p_2 = d_{0.35}$. The covariance matrix is

$$\mathrm{Cov}_{BAO}^{-1} = \begin{pmatrix} 30124 & -17227 \\ & 86977 \end{pmatrix},$$

and the vales of p_i are

$$p_1^{\mathrm{data}} = d_{0.2}^{\mathrm{data}} = 0.1905, \qquad p_2^{\mathrm{data}} = d_{0.35}^{\mathrm{data}} = 0.1097. \tag{16.25}$$

There are also some other methods capturing the information of BAO from the SDSS data, see [64, 77] and references therein for details.

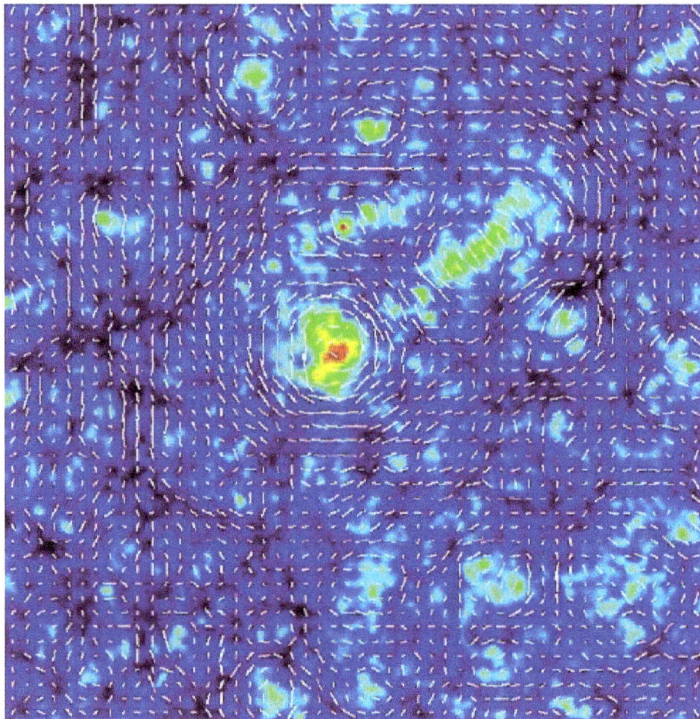

Figure 16.5　Cosmic shear field (white ticks) superimposed on the projected mass distribution from a cosmological N-body simulation: overdense regions are bright, underdense regions are dark. Note how the shear field is correlated with the foreground mass distribution. From [7]. Reprinted with kind permission of Dragan Huterer and the Annual Reviews.

16.4　Weak Lensing

Weak lensing (WL) is the slight distortions of distant galaxies' images, due to the gravitational bending of light by structures in the Universe (see Figure 16.5). Utilizing the WL effect, the distribution of dark matter and its evolution with time can be measured, thus providing a useful tool to probe dark energy through its influence on the growth of structure.

In the 1990s, WL around individual massive halos was measured [78, 79]. Soon after, WL by LSS was detected by four research groups in 2000 [80–83]. From then on, WL has grown into an increasingly accurate and powerful probe of dark matter and dark energy [84]. The current survey project of

WL is the Canada–France–Hawaii telescope legacy survey (CFHTLS) [85], covering ~ 170 square degrees.

The effect of WL on the distant sources can represent on the distortions in the shapes, sizes and brightnesses [86]. Current studies mainly focus on the change in shapes (termed as "cosmic shear"), which can be more easily and precisely measured compared with the changes in the size and brightness. The lensing effect on the shape of the galaxies is by general ~ 0.01. This effect is much smaller than the typical deviation in the galaxy shape, which is about 0.3–0.4. So a large number of galaxies are required to detect the cosmic shear signal with enough precision.

The cosmic shear analysis is a widely used method to relate the WL data with the dark energy. The cosmic shear field is the weighted mass distribution integrated along the line of sight. The Fourier transformed counterpart is the shear power spectrum P_κ. Here κ is the "convergence", which means a magnification of the source of $1 + 2\kappa$ at the linear order. P_κ can be inferred from the 3D distribution of matter, and it is related to the matter power spectrum $P(k, r)$ through

$$P_\kappa(l) = \frac{9H_0^4 \Omega_{\text{m0}}^2}{4c^4} \int_0^{r_H} \frac{dr}{a^2(r)} P\left(k = \frac{l}{f_k(r)}; r\right) \left[\int_r^{r_H} dr' n(r') \frac{f_k(r' - r)}{f_k(r')}\right].$$

(16.26)

Here $r(z) = \int_0^z \frac{dz}{H(z)}$ is the comoving distance, $n(z)$ is the mean redshift distribution of the source galaxies normalized to unity, r_H is the comoving horizon distance (i.e. the depth of the survey).

For statistical analysis of cosmic shear, it is common to use 2-point correlation functions [87]. For example, a convenient way is to describe the shear field in terms of E/B modes correlation functions [88]

$$\xi_E(\theta) = \frac{\xi_+(\theta) + \xi'(\theta)}{2}, \qquad \xi_B(\theta) = \frac{\xi_+(\theta) - \xi'(\theta)}{2}, \qquad (16.27)$$

where

$$\xi_+(\theta) = \int_0^\infty \frac{dl}{2\pi} l P_\kappa(l) J_0(l\theta), \qquad \xi_-(\theta) = \int_0^\infty \frac{dl}{2\pi} l P_\kappa(l) J_4(l\theta), \quad (16.28)$$

and

$$\xi'(\theta) = \xi_-(\theta) + 4 \int_\theta^\infty d\theta' \frac{\xi_-(\theta')}{\theta'} - 12\theta^2 \int_\theta^\infty d\theta' \frac{\xi_-(\theta')}{\theta'^3}. \qquad (16.29)$$

Here J_0 and J_4 are the zeroth and forth order Bessel functions of the first kind, respectively. The shear correlation functions can then be compared with the measurements directly. The B mode correlation functions are expected to be very small [89], so it is common to use the E mode correlation functions to construct the weak lensing χ^2 function,

$$\chi^2_{\zeta_{\mathrm{E}}} = (\zeta_{\mathrm{E}}(\theta_i) - m_i)\,(\mathrm{Cov}^{-1}_{\zeta_{\mathrm{E}}})_{ij}\,(\zeta_{\mathrm{E}}(\theta_j) - m_j)\,. \qquad (16.30)$$

Systematic errors in weak lensing measurements arise from a number of sources, including incorrect shear estimates, uncertainties in galaxy photometric redshift estimates, intrinsic correlations of galaxy shapes, and theoretical uncertainties in the mass power spectrum on small scales [90, 91]. Fortunately, future WL surveys have ability to internally constrain the impact of such effects [69, 92, 93]. For more details about the WL observation and its applications on the probe of dark energy, see [94, 86, 95] and references therein.

16.5   Galaxy Clusters

Galaxy clusters (CL) are the largest gravitationally bound objects in the universe. They typically contain 50 to 1000 galaxies and have a diameter from 2 to 10 Mpc. CL can be detected through the following approaches [96]: (i) Optical or infrared imaging and spectroscopy (see Figure 16.6). In comparison with optical surveys, infrared searches are more useful for finding higher redshift clusters. (ii) X-ray imaging and spectroscopy (see Figure 16.6). CL with active galactic nucleus (AGN) are the brightest X-ray emitting extragalactic objects, so they are quite prominent in X-ray surveys. (iii) Sunyaev–Zel'dovich effect [63]. The hot electrons in the intracluster medium scatter radiation from the cosmic microwave background through inverse Compton scattering. This produces a "shadow" in the observed CMB at some radio frequencies. (iv) Gravitational lensing. CL bend the light from distance galaxies and distort the observed images. The observed distortions can be used to detect the masses of clusters.

In principle, the number density of cluster-sized dark halos $n(z, M)$ as a function of redshift z and halo mass M can be accurately predicted from N-body simulations [98]. Comparing these predictions to large area cluster surveys can provide precise constraints on the cosmic expansion history [99–101].

In a survey that selects clusters according to some observable O with redshift-dependent selection function $f(O, z)$, the redshift distribution

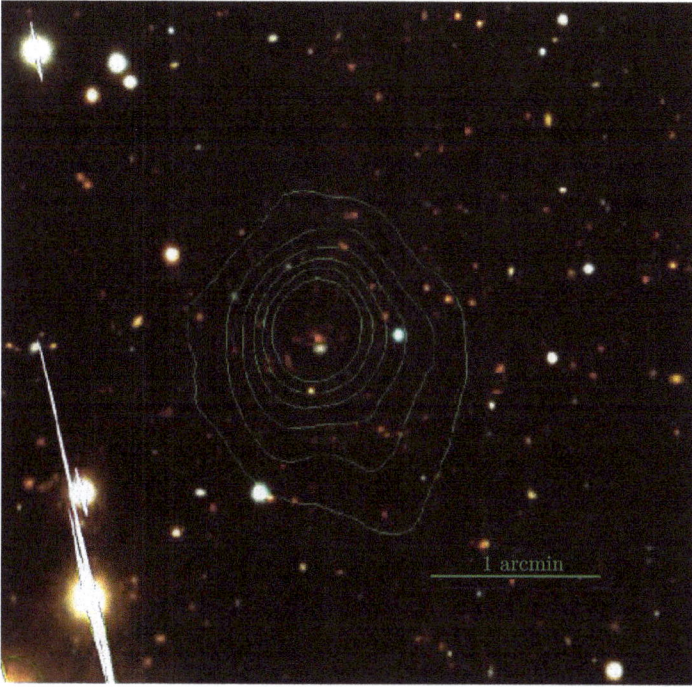

Figure 16.6   Massive galaxy cluster 2XMM J083026+524133 detected by the X-ray telescope XMM-Newton (the green contours) and its optical image observed by the Large Binocular Telescope (LBT) in the Arizona desert. Credit: [97], reproduced with permission of Georg Lamer and ©ESO.

of CL is given by [7]

$$\frac{d^2 N(z)}{dz d\Omega} = \frac{r^2(z)}{H(z)} \int_0^\infty f(O, z) dO \int_0^\infty p(O|M, z) \frac{dn(z)}{dM} dM . \qquad (16.31)$$

Here N is the number of CL, Ω is the solid angle, $r(z)$ is the comoving distance, $H(z)$ is the Hubble parameter, $dn(z)/dM$ is the space density of dark halos in comoving coordinates, and $p(O|M, z)$ is the mass-observable relation, the probability that a halo of mass M at redshift z is observed as a cluster with observable property O. The utility and errors of this probe depend on the ability to robustly associate cluster observables such as cluster galaxy richness, X-ray luminosity, Sunyaev–Zel'dovich effect flux decrement, or weak lensing shear with cluster mass [7, 96]. As seen in this equation, the sensitivity of cluster counts to dark energy arises from two factors: cosmic expansion history, $r^2(z)/H(z)$ is the comoving volume element that contains information of the cosmic expansion history; growth of structure, $dn(z)/dM$ depends on the evolution of density fluctuations, and the cluster mass function is also determined by the primordial spectrum of density perturbations. One can see [102] and references therein for more details about the surveys of CL and their applications on the probe of dark energy.

16.6   Gamma-Ray Burst

Gamma-ray bursts (GRB) are flashes of gamma rays associated with extremely energetic explosions in distant galaxies. They are the most luminous electromagnetic events in the universe.

GRBs were first detected in 1967 by the U.S. Vela satellites. They can be classified into "long GRBs" (their durations are longer than 2 seconds) and "short GRBs" (their durations are shorter than 2 seconds). At the beginning, most astronomers believed that GRBs originate from inside the Milky Way galaxy, only Paczynski insisted that GRBs originate from external galaxy [103]. In 1997, an X-ray astronomy satellite BeppoSAX detected the "afterglow" of GRB 970228 [104], and verified that GRBs indeed originate from external galaxy. This discovery opened up a new era in the history of GRBs studies.

Many satellites had been launched to probe GRBs in the past decade. The Swift mission [105] was launched in 2004. This satellite is equipped with a very sensitive gamma-ray detector as well as on-board X-ray and optical telescopes, which can be rapidly and automatically slewed to observe afterglow emission following a burst. In 2008, the Fermi mission [106] was launched. Its main instruments include the Large Area Telescope (LAT) and the Gamma-Ray Burst Monitor (GBM), the former is used to perform an all-sky survey, and the latter is used to detect sudden flares of gamma-rays. Meanwhile, on the ground, numerous optical telescopes have been built or modified to incorporate robotic control software that responds immediately to signals sent through the Gamma-ray Burst Coordinates Network (GCN) [107]. This allows telescopes to rapidly repoint towards a GRB, often within seconds of receiving the signal and while the gamma-ray emission itself is still ongoing.

GRBs have been proposed to be a complementary probe to the SNIa in the studies of dark energy. For example, in [108], Schaefer presented

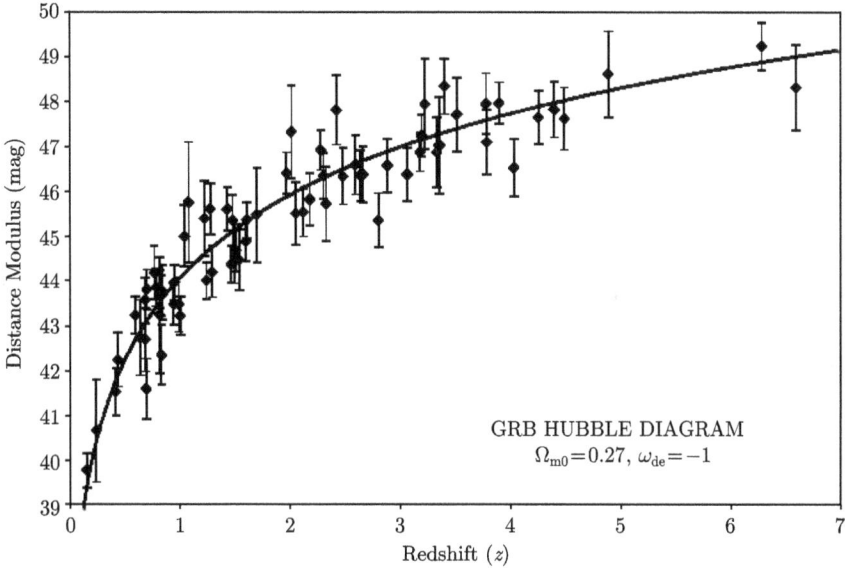

Figure 16.7   The Hubble diagram for 69 GRBs, out to the redshift of $z > 6$. The curve is the luminosity distance in a flat ΛCDM model with $\Omega_{m0} = 0.27$ and $w_{de} = -1$. It can be seen that the observational data smoothly follows the curve. From [108]. Reproduced by permission of Bradley Schaefer and the AAS.

69 GRBs data points (see Figure 16.7) over a redshift range of $z = 0.17$ to $z > 6$ with half the bursts having a redshift larger than 1.7, and showed that the GRB Hubble diagram is consistent with the existence of cosmological constant. Compared to the SNIa, an advantage of GRB is that the high energy photons in the gamma-ray band are almost immune to dust extinction. Moreover, the redshifts of observed GRBs are much higher than SNIa: there have been many GRBs observed at $1 \leqslant z \leqslant 8$, whereas the maximum redshift of GRBs is expected to be 10 or even larger [109]. Therefore, GRBs are considered to be a promising probe to fill the redshift desert between the redshifts of SNIa and CMB. For more details on the GRB cosmology, see [110, 111] and references therein.

However, there still exist some troubles in the application of GRB data to the probe of dark energy. A big problem is that since our knowledge on the mechanisms underlying the GRB emission is still limited, treating them as standard candles is still suspicious. Another well-known problem is the so-called "circularity" problem in the GRB calibration [111, 112], mainly due to the lack of low-redshift GRBs at $z < 0.1$ which are cosmology-independent. To alleviate this problem, various statistical methods have been proposed. For examples, in [113], Wang summarized the GRB data by a set of model-independent distance measurements and provided a convenient method to use the GRB data in cosmological analysis. In [114], Liang *et al.* proposed a new method to estimate the distance modulus of GRBs by interpolating from the Hubble diagram of SNIa. Using this method, in [115], Wei obtained 59 calibrated high-redshift GRBs from a total number of 109 long GRBs [108, 116] and the Union2 SNIa sample [22]. For more statistical methods to calibrate the GRB data, see e.g. [111, 112, 114]. Due to the lack of a large amount of well observed GRBs, the current GRB data are still not able to provide forceful constraint on dark energy. Therefore, although the GRBs are considered to be a promoting probe to fill the redshift desert between SNIa and CMB, there is still a long way to use them extensively and reliably to probe dark energy.

16.7   X-Ray Observations

X-ray is an important observational branch of astronomy which deals with the study of X-ray emission from celestial objects. In 1962, utilizing a US Army V-2 rocket, Giacconi and colleagues firstly discovered cosmic X-ray source. In addition, they also discovered the existence of astronomical X-ray background [117]. Due to his great contributions to this field, Giacconi received the Nobel prize in physics in 2002.

Since Giacconi's great discovery, many X-ray satellites have been launched. The first orbiting X-ray astronomy satellite, the Uhuru satellite [118], was launched in 1970. Soon after, in 1978, the first fully imaging X-ray telescope, the Einstein Observatory [119], was launched. In 1999, two important X-ray satellites, the Chandra observatory [120] and the XMM-Newton observatory [121], were launched. The Chandra observatory has high space resolution (less than 1 arc-second) and a wide wave band (0.1–1 keV), while the XMM-Newton observatory has very high spectrum resolution.

In galaxy clusters, the X-ray emitting intracluster gas is the dominant component of baryonic mass content (it exceeds the mass of optically luminous material by a factor ~ 6 [122]). In addition, galaxy clusters also have the dark matter content [123, 124]. So the measurements of universal baryonic mass function in the clusters, $f_b = \Omega_{b0}/\Omega_{m0}$, have been widely used to determine the cosmic matter fraction Ω_{m0} [125]. In late 1990s,

Sasaki [126] and Pen [127] proposed that the dependence of f_b on the angular diameter distances to the clusters can also be used to constrain the geometry of the universe. In 2002, Allen *et al.* [128–130] carried out such a test and obtained $\Omega_{m0} = 0.30 \pm 0.04$ using a small sample of X-ray luminous galaxy clusters. In 2008, based on an improved analysis using the data of 42 clusters from Chandra X-ray observations, Allen *et al.* [131] found a detection of dark energy at 4σ CL, with $\Omega_{de0} = 0.86 \pm 0.21$ for a non-flat ΛCDM model (see Figure 16.8).

The results of [128–131] were obtained from the measurements of the apparent evolution of the cluster X-ray gas mass fraction, $f_{gas} = M_{gas}/M_{tot}$. As showed in [131], the χ^2 function for f_{gas} data can be calculated as

$$\chi^2_{f_{gas}} = \sum_{i=1}^{42} \frac{[f_{gas}^{\Lambda CDM}(z_i) - f_{gas,i}]^2}{\sigma^2_{gas,i}}, \qquad (16.32)$$

where the fitting formula of $f_{gas}^{\Lambda CDM}(z)$ is given by

$$f_{gas}^{\Lambda CDM}(z) = \frac{KA\gamma b(z)}{1 + s(z)} \left(\frac{\Omega_{b0}}{\Omega_{m0}}\right) \left[\frac{d_A^{\Lambda CDM}}{D_A(z)}\right]^{1.5}. \qquad (16.33)$$

Here D_A and $D_A^{\Lambda CDM}$ are the angular diameter distances to the clusters in the current test model and reference cosmologies, respectively. The factor $b(z) = b_0(1 + \alpha_b z)$ is a "biased factor" with $0.65 < b_0 < 1.0, -0.1 < \alpha_b < 0.1$. Parameter $s(z) = s_0(1 + \alpha_s z)$ models the baryonic mass fraction in stars, and $s_0 = (0.16 \pm 0.05)h_{70}^{0.5}, -0.2 < \alpha_s < 0.2$. γ models non-thermal pressure support in the clusters $(1.0 < \gamma < 1.1)$. K is a "calibration" constant arises from the residual uncertainty in the accuracy of the instrument calibration and X-ray modeling $(K = 1 \pm 0.1)$. The factor A is

$$A = \left(\frac{\theta_{2500}^{\Lambda CDM}}{\theta_{2500}}\right)^{\eta} \approx \left(\frac{H(z)D_A(z)}{[H(z)D_A(z)]^{\Lambda CDM}}\right)^{\eta}, \qquad (16.34)$$

where $\eta = 0.214 \pm 0.022$. One can refer to [131] for details about the origins of the parameters $b(z), s(z), K$, and A.

Like the SNIa and BAO data, the f_{gas} measurements can also probe the redshift-distance relation, with the dependence $f_{gas} \propto d_L(z)D_A(z)^{0.5}$ [131]. It has been shown that [131–133] the f_{gas} data is useful in breaking the degeneracy between some cosmological parameters such as Ω_{m0}, w_{de} and H_0 when combined with other cosmological observations. In addition to

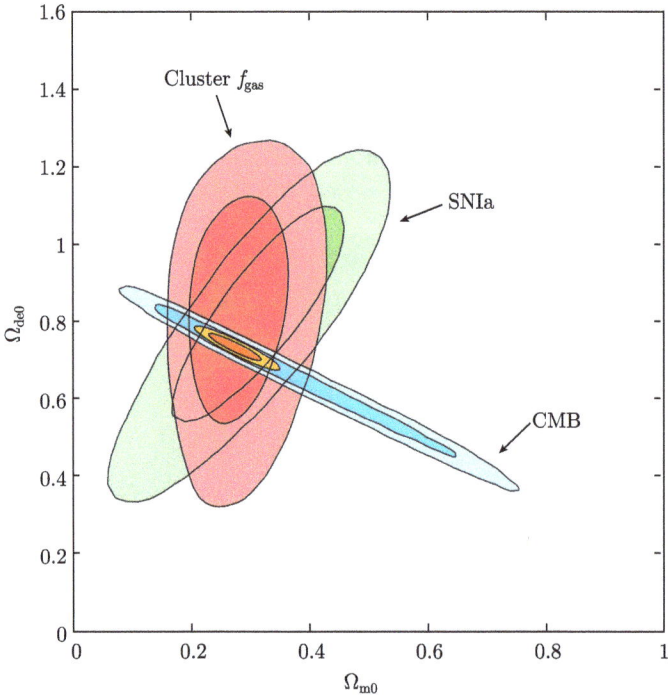

Figure 16.8   The 68.3% and 95.4% confidence constraints in the Ω_{m0}-Ω_{de0}, using Chandra measurements of the X-ray gas mass fraction f_{gas} data (red contours). A non-flat ΛCDM model is assumed. Also shown are the independent results obtained from the CMB data (blue contours) and SNIa data (green contours). A combined analysis using all the three data sets yields a result of $\Omega_{de0} = 0.735 \pm 0.023$ and $\Omega_{k0} = -0.010 \pm 0.011$ at the 68.3% CL (the inner, orange contours). From [131]. Used with permission of Steven Allen and Wiley. Copyright (2008) Wiley.

f_{gas}, measurements of the amplitude and evolution of matter fluctuations using X-ray observations have also been applied to probe dark energy. One can refer to [129, 134] for more details.

16.8   Hubble Parameter Measurements

In 1929, Hubble [135] discovered a linear correlation between the apparent distances to galaxies D and their recessional velocities v:

$$v = H_0 D \,, \tag{16.35}$$

where H_0 is the so-called Hubble constant. This discovery provided strong evidence that our universe is in a state of expansion and opened the era of the modern cosmology. Soon after, people find that this Hubble's law is just an approximate formula, and H_0 should be replaced by $H(z)$, a function of the redshift z. As mentioned above, $H(z)$ describes the expansion history of the universe, and plays a central role in connecting dark energy theories and observations.

At the beginning, Hubble and Humason [136] measured a value for H_0 of 500 km/s/Mpc, which is much higher than the currently accepted value due to errors in their distance calibrations. Since the launch of the Hubble Space Telescope (HST) [137], the value of H_0 was estimated to be between 50 and 100 km/s/Mpc. For example, using the HST key project, Freedman et $al.$ [138] obtained $H_0 = 72 \pm 8$ km/s/Mpc, and Sandage et $al.$ [139] advocated a lower $H_0 = 62.3 \pm 6.3$ km/s/Mpc. In 2009, Riess et $al.$ [140] gave $H_0 = 74.2 \pm 3.6$ km/s/Mpc with a 4.8% uncertainty. In a latest work [141], Riess et $al.$ obtained $H_0 = 73.8 \pm 2.4$ km/s/Mpc, corresponding to a 3.3% uncertainty. In the future with more precise observations from HST, Spitzer [142], Global Astrometric Interferometer for Astrophysics satellite (GAIA) [143] and James Webb Space Telescope (JWST) [144], an uncertainty of 1% in the Hubble constant will be a realistic goal for the next decade [145].

The precise measurements of H_0 will be helpful to break the degeneracy between some cosmological parameters [145]. For example, in [146], Hu pointed out that a measurement of H_0 to the percent level, when combined with CMB measurements with the statistical precision of the Planck satellite, offers one of the most precise measurements of dark energy EOS at $z \sim 0.5$.

In addition to the direct measurements for the H_0, the precise measurements of $H(z)$ are also useful in studying the cosmic acceleration. In 2005, Simon et $al.$ [147] measured the Hubble parameter $H(z)$ at nine different redshifts from the differential ages of passively evolving galaxies. In 2009, Stern et $al.$ [148] extended this dataset to eleven data points (see Table 16.2). Soon after, by studying the clustering of LRG galaxies in the

Table 16.2   Values of H_0 Measured in [147, 148]

z	0.1	0.17	0.27	0.4	0.48	0.88	0.9	1.3	1.43	1.53	1.75
$H(z)$	69	83	77	95	97	90	117	168	177	140	202
σ_H	12	8	14	17	60	40	23	17	18	14	40

Table 16.3   Values of H_0 Measured in [149]

z	0.24	0.34	0.43
$H(z)$	79.69	83.80	86.45
$\sigma_{H,\text{st}}$	2.32	2.96	3.27
$\sigma_{H,\text{sys}}$	1.29	1.59	1.69

The inferred $H(z)$ with its statistical and systematical errors for each redshift slice.

latest spectroscopic SDSS data releases, Gaztañaga *et al.* [149] obtained three more $H(z)$ data points (see Table 16.3). Utilizing these $H(z)$ data, it is straightforward to put constraint on dark energy parameters by calculating the corresponding χ^2 as

$$\chi_H^2 = \sum_{i=1}^{14} \frac{[H_{\text{obs}}(z_i) - H(z_i)]^2}{\sigma_{\text{hi}}^2} . \tag{16.36}$$

As shown in [150], the current $H(z)$ data from direct measurements can provide valuable constraints on dark energy. In the future, with the developments in the observational technique of LRGs, the $H(z)$ measurements can provide useful complements to other cosmic observations [151, 152].

16.9   Cosmic Age Tests

The cosmic age problem is a longstanding issue in cosmology and provides an important tool for constraining the expanding history of the Universe [153]. Before the great discovery that our universe is undergoing an accelerated expansion, the most popular SCDM model (i.e., a flat universe with $\Omega_m = 1$) is always plagued by a longstanding puzzle: in this model the present age of the universe is $t_0 = \frac{2}{3H_0} \simeq 9$ Gyr, while astronomers have already discovered that many objects are older than 10 Gyr [154, 155]. For example, based on the study of white dwarf cooling, a lower limit of cosmic age $t_0 = 12.7 \pm 0.7$ Gyr has been obtained [156]. The cosmic age problem becomes more acute if one considers the age of the universe at a high redshift. For instance, a 3.5 Gyr-old galaxy 53W091 at redshift $z = 1.55$ and a 4 Gyr-old galaxy 53W069 at $z = 1.43$ are more difficult to accommodate in the SCDM model [157]. Along with the discovery of accelerated expansion of the universe and the return of the cosmological constant Λ, the cosmic age problem has been greatly alleviated. The 7-year

WMAP observations show that in the ΛCDM model the present cosmic age is $t_0^{\mathrm{obs}} = 13.75 \pm 0.11$ Gyr. Besides, it is shown that the ΛCDM model can also easily accommodate galaxies 53W091 and 53W069 [158]. So the cosmic age problem is a "smoking-gun" of evidence for the existence of dark energy.

However, the cosmic age problem has not been completely removed by the introduction of dark energy. By comparing photometric data acquired from the Beijing–Arizona–Taiwan–Connecticut system with up-to-date theoretical synthesis models, Ma et $al.$ [159] obtained the ages of 139 globular clusters (GCs) in the M31 galaxy, in which 9 extremely old GCs are older than the present cosmic age predicted by the 7-year WMAP observations [160]. In addition, the existence of high-z quasar APM 08279+5255 at $z = 3.91$ [161] is also a mystery [162–164]. Using the maximum likelihood values of the 7-year WMAP observations $\Omega_{m0} = 0.272$ and $h = 0.704$, the ΛCDM model can only give a cosmic age $t = 1.63$ Gyr at redshift $z = 3.91$, while the lower limit of this quasar's age is 1.8 Gyr. To accommodate these anomalous objects, some authors suggested that a lower H_0 should be advocated [139, 165], while some authors suggested that more complicated cosmological model should be taken into account [166, 167]. Therefore, the cosmic age puzzle still remains in the standard cosmology.

In addition, one can also use the ages of old galaxies to perform the best-fit analysis on dark energy. In [147], Simon et $al.$ established these so-called "lookback time-redshift" (LT) data by estimating the age of 32 old passive galaxies distributed over the redshift interval $0.11 \leqslant z \leqslant 1.84$ and the total age of the universe t_0^{obs}. The galaxy samples of passively evolving galaxies are selected with high-quality spectroscopy, and the method used to determine ages of galaxy samples indicates that systematics are not a serious source of error for these high redshift galaxies [147]. Utilizing these LT data, one can calculate the corresponding χ^2 as

$$\chi^2_{\mathrm{age}}(\mathbf{p}) = \sum_{i=1}^{32} \frac{[t^{\mathrm{obs}}(z_i; \tau) - t(z_i; \mathbf{p})]^2}{\sigma_T^2} + \frac{[t_0^{\mathrm{obs}} - t_0(\mathbf{p})]^2}{\sigma_{t_0^{\mathrm{obs}}}^2}, \qquad (16.37)$$

where $t(z; \mathbf{p})$ is the age of the universe at redshift z, given by

$$t(z; \mathbf{p}) = \int_z^\infty \frac{d\tilde{z}}{(1 + \tilde{z})H(\tilde{z}; \mathbf{p})}. \qquad (16.38)$$

Here $t_0(\mathbf{p}) = t(0; \mathbf{p})$ is the present age of the universe. $\sigma_T^2 \equiv \sigma_i^2 + \sigma_{t_0^{\mathrm{obs}}}^2$, where σ_i is the uncertainty in the individual lookback time to the ith galaxy of the sample, $\sigma_{t_0^{\mathrm{obs}}} = 0.7$ Gyr stands for the uncertainty in the total expansion age of the universe (t_0^{obs}), and τ means the time from Big Bang to the formation of the object.

16.10   Growth Factor

In addition to the expansion history of the universe, the growth of large-scale structure can also provide important constraints on dark energy and modified gravity. The growth rate of large scale structure is derived from matter density perturbation $\delta = \delta\rho_{\mathrm{m}}/\rho_{\mathrm{m}}$ in the linear regime that satisfies the simple differential equation [168]

$$\ddot{\delta} + 2H\dot{\delta} - 4\pi G_{\mathrm{eff}}\rho_{\mathrm{m}}\delta = 0 \,, \tag{16.39}$$

where the effect of dark energy is introduced through the "Hubble damping" term $2H\dot{\delta}$, and the effect of modified gravity is introduced via the effective gravitational "constant" G_{eff}. Equation (16.39) can be written in terms of the logarithmic growth factor $f = d\ln\delta/d\ln a$ [169]

$$\frac{df}{d\ln a} + f^2 + \left(\frac{\dot{H}}{H^2} + 2\right)f = \frac{3}{2}\frac{G_{\mathrm{eff}}}{G}\Omega_{\mathrm{m}}(z) \,. \tag{16.40}$$

There is no analytical solution to the Eq. (16.40), and much efforts have been paid to solve this equation [170–172]. The growth function f was shown to be well approximated by the ansatz [100, 173, 174]

$$f = \Omega_{\mathrm{m}}^\gamma(z) \,, \tag{16.41}$$

where γ is the so-called "growth index". Based on this ansatz, the approximate expressions of f for various gravitational theory have been obtained [175, 176]. Besides, there has been much interest in exploring parameterization of the growth index as a function of the redshift [177].

In Table 16.4, we list the growth data that were converted in the work of [178–180], from either measurement of redshift distorsion parameter or from various power spectrum amplitudes from Lyman-α Forest data. The list of the respective original references is also given in the Table 16.4. A caveat in using this data to constrain other cosmological models is that in various steps in the process of analysing or converting the data, the ΛCDM model was assumed. So if one wants to use the data to constrain other models, and in particular modified gravity models, one than should redo all the steps assuming that model, starting from original observations. For more studies concerning the growth factor and its application on the probe of dark energy, see e.g. [178, 179, 186, 187].

Table 16.4   The Growth Factor Data

z	f_{obs}	References
0.15	0.49 ± 0.1	[180]
0.35	0.7 ± 0.18	[74]
0.55	0.75 ± 0.18	[181]
0.77	0.91 ± 0.36	[180]
1.4	0.9 ± 0.24	[182]
3.0	1.46 ± 0.29	[183]
2.125–2.72	0.74 ± 0.24	[184]
2.2–3	0.99 ± 1.16	[185]
2.4–3.2	1.13 ± 1.16	[185]
2.6–3.4	0.99 ± 1.16	[185]
2.8–3.6	0.99 ± 1.16	[185]
3.0–3.8	0.99 ± 1.16	[185]

16.11   Other Cosmological Probes

Sandage–Loeb test [188, 189] is a method which directly measures the evolution of the universe. The idea is to measure the drifts of the comoving cosmological sources caused by the cosmic acceleration/deceleration. For an object at redshift z_s, after a period Δt_0 (t_0 stands for the cosmic age at our position), there would be small variation of its redshift due to the drift caused by the evolution of the universe,

$$\Delta z \approx \left[\frac{\dot{a}(t_0) - \dot{a}(t_s)}{a(t_s)} \right] \Delta t_0 . \tag{16.42}$$

The variations of redshifts can be obtained by direct measurements of the quasar Lyman-α absorption lines at sufficiently separated epochs (e.g., 10–30 yrs). The Sandage–Loeb test is unique in its coverage of the "redshift desert" at $2 \leqslant z \leqslant 5$, where other dark energy probes are unable to provide useful information about the cosmic expansion history. Thus, this method is expected to be a good complementary to other dark energy probes [190].

Gravitational waves (GW) observations also have potential to make interesting contributions to the studies of dark energy. In 1986, Schutz [191] found that the luminosity distance of the binary neutron stars (or black holes) can be independently determined by observing the gravitational waves generated by these systems. If their redshifts can be determined, then they could be used to probe dark energy through the Hubble diagram [192]. This so-called GW "standard sires" has drawn a lot of attentions [193]. Besides, dark energy can also leave characteristic features on the spectrum of primordial gravitational waves [194], which may be detected by future ground-based and space-based GW detectors [195, 196].

In addition, through astronomical observations, the validity of general relativity can be tested from observations of solar systems, BBN, CMB, LSS, gravitational waves, and so on. In the following we will present a brief introduction to the tests coming from the solar system and BBN. For more details about the tests of gravity theories, see [197–199] and the references therein.

Solar system tests of alternate theories of gravity are commonly described by the Parametrized Post-Newtonian (PPN) formalism [200]. In this formalism, the metric is written as a perturbation about the Minkowski metric,

$$ds^2 = -(1 + 2\Psi - 2\beta\Psi^2)dt^2 + (1 - 2\gamma\Psi)d\vec{x}^2 \,, \qquad (16.43)$$

here the potential $\Psi = -GM/r$ for the Schwarzschild metric. The parameter β stands for the nonlinearity in the superposition law of gravity, and γ describes the spacetime curvature induced by a unit mass. The parameter γ is the most relevant PPN parameter for the modified gravity theories of interest. So far, γ has been tightly constrained from various observational methods. The tightest constraint comes from time-delay measurements in the solar system, specifically the Doppler tracking of the Cassini spacecraft, which gives $\gamma - 1 = (2.1 \pm 2.3) \times 10^{-5}$ [201]. Other tests include the observed perihelion shift of Mercury's obit [202] and light deflection measurement [203], which can constrain γ at the 10^{-3} and 10^{-4} level, respectively (see Figure 16.9).

Moreover, solar system observations can also test the modified gravity models through their violations of the Strong Equivalence Principle (SEP). That will result in a difference in the free-fall acceleration of the earth and the moon towards the sun, named as the Nordtvedt effect [202]. This effect is detectable in the Lunar Laser Ranging (LLR). Current LLR data have constrained PPN deviations from the Einstein gravity at the 10^{-4} level [197]. In the near future, the bound is expected to be improved by an order of magnitude by the Apache Point Observatory for Lunar Laser-ranging Operation (APOLLO) project [204].

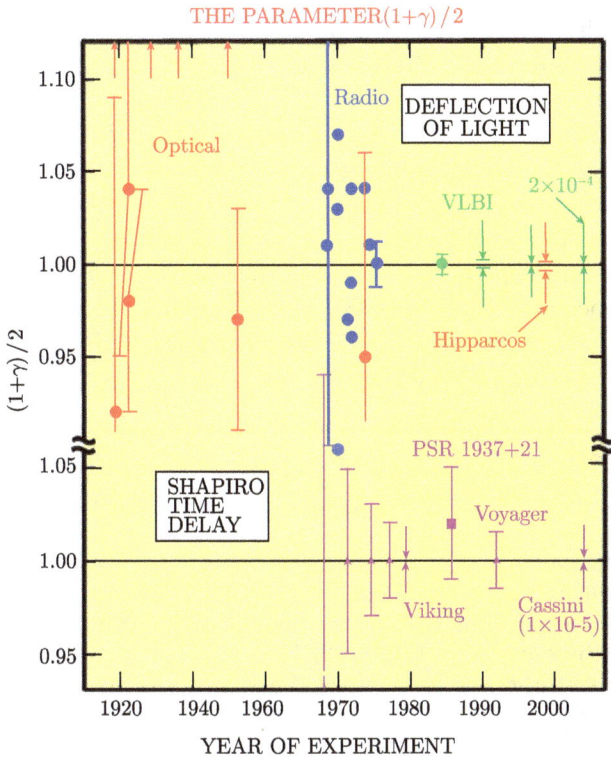

Figure 16.9   Measurements of the coefficient $(1 + \gamma)/2$ from light deflection and time delay measurements. Its value in the Einstein gravity is unity. The arrows at the top denote anomalously large values from early eclipse expeditions. The Shapiro time-delay measurements using the Cassini spacecraft yielded on agreement with Einstein gravity to 10^{-3} percent, and VLBI (very-long-baseline radio interferometry) light deflection measurements have reached 0.02 percent. Hipparcos denotes the optical astrometry satellite, which reached 0.1 percent. Figure reprinted by permission from [197]. Copyright ©Max Planck Society and Clifford M. Will.

The BBN happened when the universe was ~ 10–100 seconds old. The abundance of light elements produced during BBN is very sensitive to the Hubble parameter H and the temperature T. On the other hand, from the Friedmann Equation H and T are related to the gravitational constant G by (assume radiation domination and zero curvature)

$$H^2 \sim G g_* T^4 \,, \tag{16.44}$$

where g_* is the number of relativistic species at BBN. Based on this relation, a constraint on G at the level of better than 10% has been inferred from BBN [205].

In addition, the influence of modified gravity on H can also affect the epoch of recombination, resulting a detectable effect on the damping of the CMB power spectrum at high l. From current CMB observations, a constraint on G at $\sim 10\%$ level has been obtained [206]. The constraint from CMB can help to break the high degeneracy of Hubble parameter H and temperature T in the BBN observations and improve the constrain to $\sim 3\%$ [206]. In the future, it is expected that the Plank satellite can constrain G at the $\sim 1.5\%$ level [206].

References

[1] W. Hillebrandt and J. C. Niemeyer, Ann. Rev. Astron. Astrophys. **38** (2000) 191.
[2] B. Leibundgut *et al.*, ApJ. **466** (1996) L21.
[3] S. Perlmutter *et al.*, ApJ. **483** (1997) 565.
[4] B. Leibundgut, Ann. Rev. Astron. Astrophys. **39** (2001) 67.
[5] A. G. Riess *et al.*, AJ. **116** (1998) 1009.
[6] S. Perlmutter *et al.*, ApJ. **517** (1999) 565.
[7] J. Frieman, M. Turner, and D. Huterer, Ann. Rev. Astron. Astrophys. **46** (2008) 385.
[8] R. P. Kirshner, arXiv:0910.0257.
[9] J. L. Tonry *et al.*, ApJ. **594** (2003) 1.
[10] B. Barris *et al.*, ApJ. **602** (2004) 571.
[11] A. G. Riess *et al.*, ApJ. **607** (2004) 665.
[12] A. G. Riess *et al.*, ApJ. **659** (2007) 98.
[13] P. Astier *et al.*, Astron. Astrophys. **447** (2006) 31; S. Baumont *et al.*, Astron. Astrophys. **491** (2008) 567.
[14] N. Regnault *et al.*, arXiv:0908.3808; J. Guy *et al.*, arXiv:1010.4743.
[15] G. Miknaitis *et al.*, ApJ. **666** (2007) 674; W. M. Wood-Vasey *et al.*, ApJ. **666** (2007) 694.
[16] Y. Copin *et al.*, New Astronomy Rev. **50** (2006) 436; R. A. Scalzo *et al.*, ApJ. **713** (2009) 1073.
[17] G. Folatelli *et al.*, AJ. **139** (2010) 120; G. Folatelli *et al.*, AJ. **139** (2010) 519.
[18] J. Leaman *et al.*, arXiv:1006.4611; W. D. Li *et al.*, arXiv:1006.4612; W. D. Li *et al.*, arXiv:1006.4613.
[19] J. A. Holtzman *et al.*, AJ. **136** (2008) 2306; R. Kessler *et al.*, ApJS. **185** (2009) 32.
[20] M. Kowalski *et al.*, ApJ. **686** (2008) 749.
[21] M. Hicken *et al.*, ApJ. **700** (2009) 1097; M. Hicken *et al.*, ApJ. **700** (2009) 331.

[22] R. Amanullah *et al.*, ApJ. **716** (2010) 712.

[23] N. Suzuki *et al.*, arXiv:1105.3470.

[24] P. Hoeflich, arXiv:astro-ph/0409170; T. Plewa, A. C. Calder, and D. Q. Lamb, ApJL. **612**, (2004) L37; S. Jha, A. G. Riess, and R. P. Kirshner, ApJ. **659** (2007) 122.

[25] M. Hamuy *et al.*, AJ. **112** (1996) 2408.

[26] M. M. Phillips, ApJ. **413** (1993) L105.

[27] S. Perlmutter *et al.*, ApJ. **483** (1997) 565; M. Hamuy *et al.*, AJ. **109** (1995) 1669; A. G. Riess, W. H. Press, and R. P. Kirshner, ApJ. **473** (1996) 88.

[28] A. Kim, LBNL Report LBNL-56164 (2004).

[29] S. Nesseris and L. Perivolaropoulos, Phys. Rev. D **72** (2005) 123519; L. Perivolaropoulos, Phys. Rev. D **71** (2005) 063503; S. Nesseris and L. Perivolaropoulos, JCAP **0702** (2007) 025.

[30] http://supernova.lbl.gov/Union/.

[31] A. Conley *et al.*, ApJS. **192** (2011) 1.

[32] M. Sullivan *et al.*, arXiv:1104.1444.

[33] https://tspace.library.utoronto.ca/handle/1807/24512.

[34] Y. Wang and P. Mukherjee, ApJ. **606** (2004) 654; Y. Wang and M. Tegmark, Phys. Rev. D **71** (2005) 103513.

[35] E. V. Linder, Rept. Prog. Phys. **71** (2008) 056901.

[36] S. Perlmutter *et al.*, Bull. Am. Astron. Soc. **29** (1997) 1351; B. P. Schmidt, ApJ. **607** (1998) 46; P. M. Garnavich *et al.*, ApJ. **509** (1998) 74; S. Perlmutter, M. S. Turner, and M. J. White, Phys. Rev. Lett. **83** (1999) 670; A. G. Riess, Publ. Astron. Soc. Pac. **112** (2000) 1284; A. G. Riess *et al.*, ApJ. **560** (2001) 49; P. Nugent, A. Kim, and S. Perlmutter, Publ. Astron. Soc. Pac. **114** (2002) 803; R. A. Knop *et al.*, ApJ. **598** (2003) 102; S. Perlmutter and B. P. Schmidt, Lect. Notes Phys. **598** (2003) 195; A. V. Filippenko, arXiv:astro-ph/0410609; C. Shapiro and M. S. Turner, ApJ. **649** (2006) 563; B. Leibundgut, Gen. Rel. Grav. **40** (2008) 221.

[37] A. A. Penzias and R. W. Wilson, ApJ. **142** (1965) 419.

[38] R. A. Alpher, H. Bethe, and G. Gamow, Phys. Rev. D **73** (1948) 803.

[39] G. F. Smoot *et al.*, ApJ. **396** (1992) L1; C. L. Bennett *et al.*, ApJ. **396** (1992) L7.

[40] A. D. Miller *et al.*, ApJL. **524** (1999) L1; P. de Bernardlis *et al.*, Nature (London) **404** (2000) 955; S. Hanany *et al.*, ApJL. **545** (2000) L5.

[41] P. J. E. Peebles and J. T. Yu, ApJ. **162** (1970) 815.

[42] J. R. Bond and G. Efstathiou, ApJ. **285** (1984) L45.

[43] J. A. Holtzman, ApJS. **71** (1989) 1.

[44] W. Hu and N. Sugiyama, ApJ. **444** (1995) 489; U. Seljak and M. Zaldarriaga, ApJ. **469** (1996) 437; M. Zaldarriaga, U. Seljak, and E. Bertschinger, ApJ. **494** (1998) 491; M. Zaldarriaga and U. Seljak, ApJS. **129** (2000) 431; A. Lewis, A. Challinor, and A. Lasenby, ApJ. **538** (2000) 473; N. W. Halverson *et al.*, ApJL. **568** (2002) 38; B. S. Mason *et al.*, ApJ. **591** (2003) 549; A. T. Lee, ApJ. **561** (2001) L1; C. B. Netterfield *et al.*, ApJ. **571** (2002) 604; T. J. Pearson *et al.*, ApJ. **591** (2003) 556; S. Dodelson, *Modern Cosmology* (Academic Press, 2003).

[45] D. J. Eisenstein and W. Hu, ApJ. **496** (1998) 605.
[46] W. Hu and N. Sugiyama, ApJ. **471** (1996) 542.
[47] J. R. Bond, G. Efstathiou, and M. Tegmark, MNRAS **291** (1997) L33.
[48] C. L. Bennett *et al.*, ApJS. **148** (2003) 1; D. N. Spergel *et al.*, ApJS. **148** (2003) 175; D. N. Spergel *et al.*, ApJS. **170** (2007) 377; L. Page *et al.*, ApJS. **170** (2007) 335.
[49] E. Komatsu *et al.*, ApJS. **180** (2009) 330.
[50] C. L. Reichardt *et al.*, ApJ. **694** (2009) 1200.
[51] M. L. Brown *et al.*, ApJ. **705** (2009) 978.
[52] E. Komatsu *et al.*, arXiv:1001.4538.
[53] Planck Collaboration, arXiv:1101.2022; arXiv:1101.2023; arXiv:1101.2024.
[54] R. K. Sachs and A. M. Wolfe, ApJ. **147** (1967) 73.
[55] K. Coble, S. Dodelson, and J. A. Frieman, Phys. Rev. D **55** (1997) 1851; R. Scranton *et al.*, arXiv:astro-ph/0307335; P. S. Corasaniti *et al.*, Phys. Rev. Lett. **90** (2003) 091303; P. S. Corasaniti, T. Giannantonio, and A. Melchiorri, Phys. Rev. D **71** (2005) 123521; W. Hu *et al.*, Phys. Rev. D **57** (1998) 3290; P. Fosalba, E. Gaztanaga, and F. Castander, ApJ. **597** (2003) L89; A. Rassat *et al.*, MNRAS **377** (2007) 1085.
[56] R. R. Caldwell, R. Dave, and P. J. Steinhardt, Phys. Rev. Lett. **80** (1998) 1582.
[57] T. Giannantonio *et al.*, Phys. Rev. D **77** (2008) 123520.
[58] S. Ho *et al.*, Phys. Rev. D **78** (2008) 043519.
[59] G. Jungman *et al.*, Phys. Rev. D **54** (1996) 1332; A. H. Jaffe *et al.*, Phys. Rev. Lett. **86** (2001) 3475; E. Komatsu and D. N. Spergel, Phys. Rev. D **63** (2001) 063002; R. Stompor *et al.*, ApJ. **561** (2001) L7; J. Kovac *et al.*, Nature **420** (2002) 772; A. Benoit *et al.*, Astron. Astrophys. **399** (2003) L25; A. Kogut *et al.*, ApJS. **148** (2003) 161; K. M. Gorski *et al.*, ApJ. **622** (2005) 759; U. Seljak, A. Slosar, and P. McDonald, JCAP **0610** (2006) 014; J. Dunkley *et al.*, ApJS. **180** (2009) 306; H. Liu and T. P. Li, arXiv:0907.2731; H. Liu and T. P. Li, arXiv:1003.1073.
[60] S. Dodelson and L. Knox, Phys. Rev. Lett. **84** (2000) 3523; P. S. Corasaniti *et al.*, Phys. Rev. D **70** (2004) 083006; R. Bean and O. Dore, Phys. Rev. D **69** (2004) 083503; N. Afshordi, Y. S. Loh and M. A. Strauss, Phys. Rev. D **69** (2004) 083524; H. K. Jassal, J. S. Bagla, and T. Padmanabhan, MNRAS **405** (2010) 2639.
[61] M. Kamionkowski and A. Kosowsky, Ann. Rev. Nucl. Part. Sci. **49** (1999) 77; W. Hu and S. Dodelson, Ann. Rev. Astron. Astrophys. **40** (2002) 171; X. H. Fan, C. L. Carilli, and B. G. Keating, Ann. Rev. Astron. Astrophys. **44** (2006) 415; D. Samtleben, S. Staggs, and B. Winstein, Ann. Rev. Nucl. Part. Sci. **57** (2007) 245; M. Tristram and K. Ganga, Rept. Prog. Phys. **70** (2007) 899; N. Aghanim, S. Majumdar, and J. Silk, Rept. Prog. Phys. **71** (2008) 066902.
[62] J. Silk, ApJ. **151** (1968) 459.
[63] R. A. Sunyaev, Y. B. Zel'dovich, and B. Ya., Astrophys. & Space Science **7** (1970) 3; P. J. Zhang and U. L. Pen, ApJ. **577** (2002) 555.

[64] D. J. Eisenstein *et al.*, ApJ. **633** (2005) 560.

[65] S. Dodelson, *Modern Cosmology* (Academic Press, San Francisco, 2003).

[66] D. M. Goldberg and M. A. Strauss, ApJ. **495** (1998) 29; A. Meiksin, M. White, and J. A. Peacock, MNRAS **304** (1999) 851; V. Springel *et al.*, Nature **435** (2005) 485; H. J. Seo and D. J. Eisenstein, ApJ. **633** (2005) 575; M. White, Astroparticle Phys. **24** (2005) 334; D. J. Eisenstein, H. J. Seo, and M. White, ApJ. **664** (2007) 660.

[67] J. S. B. Wyithe and A. Loeb, ApJ. **588** (2003) L59; R. Cen, ApJ. **591** (2003) L5; Z. Haiman and G. P. Holder, ApJ. **595** (2003) 1; M. Zaldarriaga, S. R. Furlanetto, and L. Hernquist, ApJ. **608** (2004) 622; S. Furlanetto, S. P. Oh, and F. Briggs, Phys. Rept. **433** (2006) 181; M. McQuinn *et al.*, MNRAS **377** (2007) 1043; M. McQuinn *et al.*, ApJ. **653** (2006) 815; J. R. Pritchard and A. Loeb, Phys. Rev. D **78** (2008) 103511; X. C. Mao and X. Wu, ApJ. **673** (2008) L107; X. L. Chen and J. Miralda-Escude, ApJ. **684** (2008) 18.

[68] A. Cooray *et al.*, ApJ. **557** (2001) L7; H. J. Seo and D. J. Eisenstein, ApJ. **598** (2003) 720; C. Blake and K. Glazebrook, ApJ. **594** (2003) 665; W. Hu and Z. Haiman, Phys. Rev. D **68** (2003) 3004; T. Matsubara, ApJ. **615** (2004) 573; B. A. Bassett and R. Hlozek, arXiv:0910.5224.

[69] A. Albrecht *et al.*, arXiv:astro-ph/0609591.

[70] J. Guzik, G. Bernstein, and R. E. Smith, MNRAS **375** (2007) 1329; H. J. Seo and D. J. Eisenstein, ApJ. **665** (2007) 14; R. E. Smith, R. Scoccimarro and R. K. Sheth, Phys. Rev. D **77** (2008) 043525.

[71] W. J. Percival *et al.*, MNRAS **327** (2001) 1297; S. Cole *et al.*, MNRAS **362** (2005) 505; G. Huetsi, Astron. Astrophys. **449** (2006) 891.

[72] M. Colless *et al.*, arXiv:astro-ph/0306581.

[73] D. G. York *et al.*, AJ. **120** (2000) 1579.

[74] M. Tegmark *et al.*, Phys. Rev. D **74** (2006) 123507.

[75] http://www.sdss3.org/dr8/.

[76] W. J. Percival *et al.*, MNRAS **401** (2010) 2148.

[77] C. H. Chuang and Y. Wang, arXiv:1102.2251.

[78] J. A. Tyson, R. A. Wenk, and F. Valdes, ApJL. **349** (1990) L1.

[79] T. G. Brainerd, R. D. Blandford, and I. Smail, ApJ. **466** (1996) 623.

[80] D. J. Bacon, A. R. Refregier, and R. S. Ellis, MNRAS **318** (2000) 625.

[81] N. Kaiser, G. Wilson, and G. A. Luppino, arXiv:astro-ph/0003338.

[82] L. van Waerbeke *et al.*, Astron. Astrophys. **358** (2000) 30.

[83] D. M. Wittman *et al.*, Nature **405** (2000) 143.

[84] D. Huterer, Phys. Rev. D. **65** (2002) 063001.

[85] http://www.cfht.hawaii.edu/Science/CFHLS/.

[86] A. Heavens, Nucl. Phys. Proc. Suppl. **194** (2009) 76.

[87] J. Benjamin *et al.*, MNRAS **381** (2007) 702; T. Hamana, S. Colombi, and Y. Mellier, arXiv:astro-ph/0009459; L. Fu *et al.*, Astron. Astrophys. **479** (2008) 9; H. Hoekstra *et al.*, ApJ. **577** (2002) 595; H. Hoekstra *et al.*, ApJ. **647** (2006) 116.

[88] R. G. Crittenden *et al.*, ApJ. **568** (2002) 20.

[89] P. Schneider, L. Van Waerbeke, and Y. Mellier, Astron. Astrophys. **389** (2002) 729.

[90] T. D. Kitching *et al.*, MNRAS **376** (2007) 771; A. Amara and A. Refregier, MNRAS **391** (2008) 228; T. D. Kitching *et al.*, arXiv:0812.1966.
[91] D. Huterer *et al.*, MNRAS **366** (2006) 101; Z. Ma, W. Hu, and D. Huterer, ApJ. **636** (2006) 21; C. Heymans *et al.*, MNRAS **371** (2006) 750.
[92] H. Zhan, JCAP **0608** (2006) 008; H. Zhan *et al.*, ApJ. **675** (2008) L1; H. Zhan, L. Knox and J. A. Tyson, ApJ. **690** (2009) 923.
[93] A. R. Zentner, D. H. Rudd, and W. Hu, Phys. Rev. D **77** (2008) 043507.
[94] E. J. Copeland, M. Sami, and S. Tsujikawa, Int. J. Mod. Phys. D **15** (2006) 1753.
[95] Y. Mellier, Ann. Rev. Astron. Astrophys. **37** (1999) 127; M. Bartelmann and P. Schneider, Phys. Rept. **340** (2001) 291; H. Hoekstra, H. Yee, and M. Gladders, New Astron. Rev. **46** (2002) 767; T. Hamana, M. Takada, and N. Yoshidae, MNRAS **350** (2004) 893; A. Lewis and A. Challinor, Phys. Rept. **429** (2006) 1; H. Hoekstra and B. Jain, Ann. Rev. Nucl. Part. Sci. **58** (2008) 99; D. Munshi *et al.*, Phys. Rept. **462** (2008) 67; D. Huterer, Gen. Rel. Grav. **42** (2010) 2177.
[96] S. Borgani, arXiv:astro-ph/0605575.
[97] G. Lamer *et al.*, arXiv:0805.3817.
[98] M. S. Warren *et al.*, ApJ. **646** (2006) 881.
[99] Z. Haiman, J. J. Mohr, and G.P. Holder, ApJ. **553** (2001) 545.
[100] L. M. Wang and P. J. Steinhardt, ApJ. **508** (1998) 483.
[101] S. Wang *et al.*, Phys. Rev. D **70** (2004) 123008.
[102] N. A. Bahcall, Ann. Rev. Astron. Astroph. **15** (1977) 505; C. L. Sarazin, *X-Ray Emission from Clusters of Galaxies* (Cambridge University Press, 1988); P. Rosati, S. Borgani, and C. Norman, Ann. Rev. Astron. Astroph. **40** (2002) 539; A. Cooray and R. K. Sheth, Phys. Rep. **372** (2002) 1; G. M. Voit, Rev. Mod. Phys. **77** (2005) 207; S. Borgani and A. Kravtsov, arXiv:0906.4370; A. J. Benson, Phys. Rep. **495** (2010) 33; H. Bohringer and N. Werner, Ann. Rev. Astron. Astroph. **18** (2010) 127; S. W. Allen, A. E. Evrard, and A. B. Mantz, arXiv:1103.4829.
[103] B. Paczynski, ApJ. **308** (1986) L43; B. Paczynski, PASP **107** (1995) 1167.
[104] E. Costa *et al.*, Nature **387** (1997) 783; J. Paradijs *et al.*, Nature **386** (1997) 686; D. A. Frail *et al.*, Nature **389** (1997) 261.
[105] N. Gehrels *et al.*, ApJ. **611** (2004) 1005; P. Romano, arXiv:1010.2206.
[106] W. B. Atwood *et al.*, ApJ. **697** (2009) 1071.
[107] http://gcn.gsfc.nasa.gov.
[108] B. E. Schaefer, ApJ. **660** (2007) 16.
[109] V. Bromm and A. Loeb, ApJ. **575** (2002) 111; J. R. Lin, S. N. Zhang, and T. P. Li, ApJ. **605** (2004) 819.
[110] T. P. Li and Y. Q. Ma, ApJ. **272** (1983) 317; L. Amati *et al.*, Astron. Astrophys. **390** (2002) 81; G. Ghirlanda, G. Ghisellini, and D. Lazzati, ApJ. **616** (2004) 331; B. Zhang, Chin. J. Astron. Astrophys. **7** (2007) 1; B. Zhang, arXiv:astro-ph/0611774; P. Meszaros, Rept. Prog. Phys. **69** (2006) 2259; S. E. Woosley and J. S. Bloom, Ann. Rev. Astron. Astrophys. **44** (2006) 507.
[111] G. Ghirlanda, G. Ghisellini, and C. Firmani, New J. Phys. **8** (2006) 123.

[112] G. Ghirlanda *et al.*, ApJ. **613** (2004) L13; H. Li *et al.*, ApJ. **680** (2008) 92; H. Wei and S. N. Zhang, Eur. Phys. J. C **63** (2009) 139.

[113] Y. Wang, Phys. Rev. D **78** (2008) 123532.

[114] E. W. Liang and B. Zhang, MNRAS **369** (2006) L37; N. Liang *et al.*, ApJ. **685** (2008) 354.

[115] H. Wei, JCAP **1008** (2010) 020.

[116] L. Amati *et al.*, MNRAS **391** (2008) 577.

[117] R. Giacconi *et al.*, Phys. Rev. Lett. **9** (1962) 439.

[118] http://heasarc.gsfc.nasa.gov/docs/uhuru/uhuru.html.

[119] http://heasarc.gsfc.nasa.gov/docs/einstein/heao2.html.

[120] http://www.nasa.gov/mission pages/chandra/main/index.html.

[121] http://xmm.esac.esa.int/.

[122] M. Fukugita, C. J. Hogan, and P. J. E. Peebles, ApJ. **503** (1998) 518.

[123] S. D. M. White, Nature **366** (1993) 429.

[124] V. R. Eke, J. F. Navarro, and C. S. Frenk, ApJ. **503** (1998) 569.

[125] S. D. M. White and C. S. Frenk, ApJ. **379** (1991) 52; A. C. Fabian, MNRAS **253** (1991) L29; D. A. White and A. C. Fabian, MNRAS **273** (1995) 72; L. P. David, C. Jones, and W. Forman, ApJ. **445** (1995) 578; A. E. Evrard, MNRAS **292** (1997) 289; A. J. R. Sanderson and T. J. Ponman, MNRAS **345** (2003) 1241; Y. T. Lin, J. J. Mohr, and S. S. Stanford, ApJ. **591** (2003) 794; S. J. LaRoque *et al.*, ApJ. **657** (2006) 917.

[126] S. Sasaki, Publ. Astron. Soc. Jpn. **48** (1996) L119.

[127] U. Pen, New Astronomy **2** (1997) 309.

[128] S. W. Allen, R. W. Schmidt, and A. C. Fabian, MNRAS **334** (2002) L11.

[129] S. W. Allen, *et al.*, MNRAS **342** (2003) 287.

[130] S. Ettori, P. Tozzi, and P. Rosati, Astron. Astrophys. **398** (2003) 879.

[131] S. W. Allen *et al.*, MNRAS **383** (2008) 879.

[132] D. Rapetti, S. W. Allen, and J. Weller, MNRAS **360** (2005) 555.

[133] D. Rapetti and S. W. Allen, MNRAS **388** (2008) 1265.

[134] S. Borgani *et al.*, ApJ. **561** (2001) 13; T. H. Reiprich and H. Bohringer, ApJ. **567** (2002) 716; P. Schuecker *et al.*, Astron. Astrophys. **402** (2003) 53; A. Voevodkin and A. Vikhlinin, ApJ. **601** (2004) 610; J. P. Henry, ApJ. **609** (2004) 603; S. W. Allen *et al.*, MNRAS **353** (2004) 457; R. W. Schmidt and S. W. Allen, MNRAS **379** (2007) 209; A. Mantz *et al.*, MNRAS **387** (2008) 1179.

[135] E. P. Hubble, PNAS **15** (1929) 168.

[136] E. Hubble and M. L. Humason, ApJ. **74** (1931) 43.

[137] http://hubblesite.org/.

[138] W. L. Freedman *et al.*, ApJ. **533** (2001) 47.

[139] A. Sandage *et al.*, ApJ. **653** (2006) 843.

[140] A. G. Riess *et al.*, ApJ. **699** (2009) 539.

[141] A. G. Riess *et al.*, ApJ. **730** (2011) 119.

[142] http://www.spitzer.caltech.edu/.

[143] http://www.esa.int/export/esaSC/120377 index 0 m.html.

[144] http://www.jwst.nasa.gov/.

[145] W. L. Freedman and B. F. Madore, Annu. Rev. Astron. Astrophys. **48** (2010) 673.
[146] W. Hu, ASP Conf. Ser. **339** (2005) 215.
[147] J. Simon, L. Verde, and R. Jimenez, Phys. Rev. D **71** (2005) 123001.
[148] D. Stern *et al.*, JCAP **1002** (2010) 008.
[149] E. Gaztañaga, A. Cabré, and L. Hui, MNRAS **399** (2009) 1663.
[150] H. Y. Wan *et al.*, Phys. Lett. B **651** (2007) 352; Y. Gong *et al.*, arXiv:0810.3572; Z. X. Zhai, H. Y. Wan and T. J. Zhang, Phys. Lett. B **689** (2010) 8; H. R. Yu *et al.*, Research in Astronomy and Astrophysics **11** (2011) 125.
[151] H. Lin *et al.*, Mod. Phys. Lett. A **24** (2009) 1699; C. Ma and T. J. Zhang, ApJ. **730** (2011) 74; Z. X. Zhai, T. J. Zhang, and W. B. Liu, JCAP **08** (2011) 019.
[152] T. J. Zhang, C. Ma and T. Lan, Advances in Astronomy **2010** (2010) 184284.
[153] B. Chaboyer, Phys. Rept. **307** (1998) 23.
[154] H. B. Richer *et al.*, ApJ. **574** (2002) L151.
[155] L. M. Krauss and B. Chaboyer, Science **299** (2003) 65.
[156] B. M. S. Hansen *et al.*, ApJ. **574** (2002) L155.
[157] J. Dunlop *et al.*, Nature **381** (1996) 581; H. Spinrad *et al.*, ApJ. **484** (1997) 581.
[158] J. S. Alcaniz and J. A. S. Lima, ApJ. **521** (1999) L87.
[159] J. Ma *et al.*, AJ. **137** (2009) 4884; S. Wang *et al.*, AJ. **139** (2010) 1438.
[160] S. Wang, X. D. Li, and M. Li, Phys. Rev. D **82** (2010) 103006.
[161] G. Hasinger, N. Schartel, and S. Komossa, ApJ. **573** (2002) L77.
[162] A. Friaca, J. S. Alcaniz, and J. A. S. Lima, MNRAS **362** (2005) 1295; D. Jain and A. Dev, Phys. Lett. B **633** (2006) 436; N. Pires, Z. H. Zhu, and J. S. Alcaniz, Phys. Rev. D **73** (2006) 123530; H. Wei and S. N. Zhang, Phys. Rev. D **76** (2007) 063003.
[163] B. Wang *et al.*, Phys. Lett. B **662** (2008) 1.
[164] B. Wang *et al.*, Nucl. Phys. B **778** (2007) 69.
[165] G. A. Tammann *et al.*, arXiv:astro-ph/0112489.
[166] S. Wang and Y. Zhang, Phys. Lett. B **669** (2008) 201.
[167] J. L. Cui and X. Zhang, Phys. Lett. B **690** (2010) 233.
[168] P. J. E. Peebles, *The Large-Scale Structure of the Universe* (Princeton University Press, Princeton, New Jersey, 1980).
[169] S. M. Carroll, W. H. Press, and E. L. Turner, Ann. Rev. Astron. Astrophys. **30** (1992) 499.
[170] A. Lue, R. Scoccimarro, and G. Starkman, Phys. Rev. D **69** (2004) 124015.
[171] K. Koyama and R. Maartens, JCAP **0610** (2006) 016.
[172] D. Polarski and R. Gannouji, Phys. Lett. B **660** (2008) 439; Y. G. Gong, Phys. Rev. D **78** (2008) 123010; W. Hu and I. Sawicki, Phys. Rev. D **76** (2007) 104043; S. Daniel *et al.*, Phys. Rev. D **77** (2008) 103513.
[173] J. N. Fry, Phys. Lett. B **158** (1985) 211.
[174] A. P. Lightman and P. L. Schechter, ApJ. **74** (1990) 831.

[175] D. Lahav *et al.*, MNRAS **251** (1991) 128.

[176] A. J. S. Hamilton, MNRAS **322** (2001) 419.

[177] E. V. Linder and R. N. Cahn, Astropart. Phys. **28** (2007) 481; Y. G. Gong, M. Ishak, and A. Z. Wang, Phys. Rev. D **80** (2009) 023002; M. J. Mortonson, W. Hu, and D. Huterer, Phys. Rev. D **79** (2009) 023004; M. Ishak and J. Dossett, Phys. Rev. D **80** (2009) 043004; J. Dossett *et al.*, JCAP **1004** (2010) 022.

[178] C. Di Porto and L. Amendola, Phys. Rev. D **77** (2008) 083508.

[179] S. Nesseris and L. Perivolaropoulos, Phys. Rev. D **77** (2008) 023504.

[180] L. Guzzo *et al.*, Nature **451** (2008) 541.

[181] N. P. Ross *et al.*, MNRAS **381** (2007) 573.

[182] J. Da Ângela *et al.*, MNRAS **383** (2008) 565.

[183] P. McDonald *et al.*, ApJ. **635** (2005) 761.

[184] M. Viel, M. G. Haehnelt, and V. Springel, MNRAS **354** (2004) 684.

[185] M. Viel *et al.*, MNRAS **365** (2006) 231.

[186] D. Papetti *et al.*, arXiv:0911.1787.

[187] M. J. Mortonson, W. Hu and D. Huterer, Phys. Rev. D **81** (2010) 063007.

[188] A. Sandage, ApJ. **139** (1962) 319.

[189] A. Loeb, ApJ. **499** (1998) L111.

[190] P. S. Corasaniti, D. Huterer, and A. Melchiorri, Phys. Rev. D **75** (2007) 062001.

[191] B. Schutz, Nature (London) **323** (1986) 310.

[192] D. E. Holz and S. A. Hughes, ApJ. **629** (2005) 15; N. Dalal *et al.*, Phys. Rev. D **74** (2006) 063006.

[193] K. G. Arun *et al.*, Phys. Rev. D **76** (2007) 104016; C. L. MacLeod and C. J. Hogan, Phys. Rev. D **77** (2008) 043512; K. G. Arun and C. M. Will, Class. Quant. Grav. **26** (2009) 155002; C. Cutler and D. E. Holz, Phys. Rev. D **80** (2009) 104009; B. S. Sathyaprakash, B. F. Schutz, and C. Van Den Broeck, Class. Quant. Grav. **27** (2010) 215006; S. Nissanke *et al.*, ApJ. **725** (2010) 496; W. Zhao *et al.*, Phys. Rev. D **83** (2011) 023005.

[194] L. A. Boyle and P. J. Steinhardt, Phys. Rev. D **77** (2008) 063504.

[195] L. P. Grishchuk, Class. Quant. Grav. **14** (1997) 1445; L. P. Grishchuk *et al.*, Phys. Usp. **44** (2001) 1; L. P. Grishchuk, Lect. Notes Phys. **562** (2001) 167; L. P. Grishchuk *et al.*, Class. Quant. Grav. **22** (2005) 245.

[196] Y. Zhang *et al.*, Class. Quant. Grav. **22** (2005) 1383; Y. Zhang *et al.*, Class. Quant. Grav. **23** (2006) 3783; W. Zhao and Y. Zhang, Phys. Rev. D **74** (2006) 043503; H. X. Miao and Y. Zhang, Phys. Rev. D **75** (2007) 104009; S. Wang *et al.*, Phys. Rev. D **77** (2008) 104016; S. Wang, Phys. Rev. D **81** (2010) 023002.

[197] C. M. Will, Living Rev. Rel. **9** (2005) 3. URL: http://www.livingreviews.org/lrr-2006-3.

[198] B. Jain and J. Khoury, Annals Phys. **325** (2010) 1479.

[199] J. P. Uzan, Rev. Mod. Phys. **75** (2003) 403; J. P. Uzan, Living Rev. Real. **4** (2011) 2.

[200] K. Nordtvedt, Phys. Rev. **169** (1968) 1027.

[201] S. S. Shapiro et al., Phys. Rev. Lett. 92 (2004) 121101.
[202] K. Nordtvedt, Phys. Rev. 169 (1968) 1014.
[203] I. I. Shapiro, in N. Ashby, D. F. Bartlett, and W. Wyss, eds., General Relativity and Gravitation, in Proceedings of the 12th International Conference on General Relativity and Gravitation, University of Colorado at Boulder, July 2–8, 1989, 313–330 (Cambridge University Press, Cambridge, U.K., New York, U.S.A., 1990).
[204] J. G. Williams, S. G. Turyshev, and T. W. Murphy, Int. J. Mod. Phys. D 13 (2004) 567.
[205] F. Iocco et al., Phys. Rep. 472 (2009) 1.
[206] S. Galli et al., Phys. Rev. D 80 (2009) 023508.

17
Dark Energy Projects

Here we provide a brief overview of present and future projects to probe dark energy. Although some high energy physics experiments (such as LHC [1]) might shed light on dark energy through discoveries about supersymmetry or dark matter, here we only introduce the projects involving cosmological observations. According to the DETF report [2], the dark energy projects can be classified into four stages: completed projects are Stage I; on-going projects, either taking data or soon to be taking data, are Stage II; intermediate-scale, near-future projects belong to Stage III; larger-scale, longer-term future projects belong to Stage IV.

Moreover, to compare various dark energy projects, DETF [2] also proposed a quantity called figure of merit (FoM). Utilizing the famous Chevallier–Polarski–Linder (CPL) parameterization

$$w_{\mathrm{de}}(a) = w_0 + (1 - a)w_a \,, \qquad (17.1)$$

a FoM is defined as the reciprocal of the area of an error ellipse in the $w_0 - w_a$ plane. A conventional normalization takes for the FoM the square root of the determinant of the 2×2 Fisher matrix for w_0 and w_a. Soon after, an extended version of FoM was proposed by Wang [3]. More advanced stages are expected to deliver tighter dark energy constraints, and give larger FoMs. For examples, Stage III experiments are expected to deliver a factor ~ 3 improvement in the FoM compared to the combined Stage II results, while Stage IV experiments will improve the FoM by a factor of 10 compared to Stage II.

In this chapter, we will introduce some most representative projects of Stage II, Stage III and Stage IV. The corresponding dark energy projects are shown in Figure 17.1. The Stage I experiments that have already reported results, such as SNLS, ESSENCE, and WMAP, will not be discussed here.

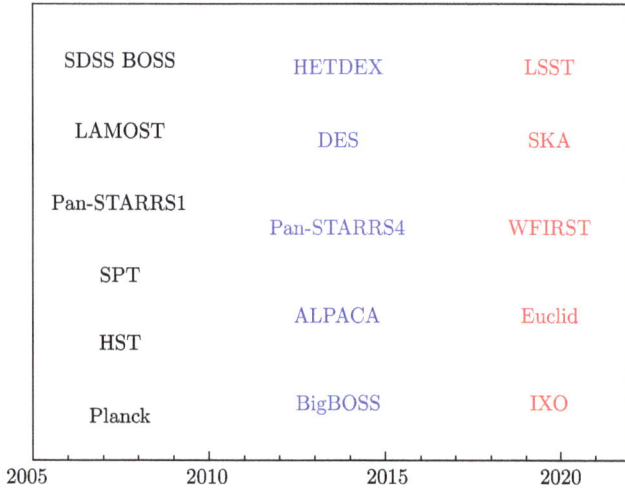

Figure 17.1   Dark energy projects introduced in this work. The black ones are stage II projects, the blue ones are stage III projects, and the red ones are stage IV projects.

Notice that some dark energy projects have already launched since 2006, our classifications of dark energy projects are slightly different from that in the DETF report. For a comprehensive list of the dark energy experiments, see [4].

17.1   On-Going Projects

In this section we introduce the on-going projects. A list of some most representative projects of Stage II is given in Table 17.1.

Table 17.1   On-Going Projects

Survey	Location	Description	Probes
SDSS BOSS	Sacramento (USA)	Optical, 2.5-m	BAO
LAMOST	Xinglong (China)	Optical, 4-m	BAO
Pan-STARRS1	Hawaii (USA)	Optical, 1.8-m	SN, WL, CL
SPT	South Pole	Submillimeter, 10-m	CL
HST	Low Earth Orbit	Optical/Near-Infrared	SN
Planck	Sun–Earth L2 Orbit	SZE	CL

The SDSS Baryon Oscillation Spectroscopic Survey (BOSS) [5] is one of the four surveys of SDSS-III, using a dedicated wide-field, 2.5-meter optical telescope at Apache Point Observatory in Sacramento Mountains. With a 5-year survey (2009–2014) of 1.5 million luminous red galaxies (LRGs), BOSS will achieve the first measurement of the BAO absolute cosmic distance scale with 1% precision at redshifts $z = 0.3$ and $z = 0.6$. It will also measure the distribution of quasar absorption lines at $z = 2.5$, yielding a measurement of the angular diameter distance at that redshift to an accuracy of 1.5%.

The Large Sky Area Multi-Object Fiber Spectroscopic Telescope (LAMOST) [6] is a National Major Scientific Project (NMSP) built by the Chinese Academy of Sciences (CAS). It is a wide-field, 4-meter ground-based optical telescope located at the Xinglong observing station of National Astronomical Observatories (NAO). The main construction of this instrument was finished in 2008. After its commissioning stage, LAMOST will study dark energy through the BAO technique.

The Panoramic Survey Telescope and Rapid Response System (Pan-STARRS) [7] is an international collaboration program led by University of Hawaii. Its first stage, Pan-STARRS1, is a 1.8-m wide-field telescope located at Haleakala in Hawaii. The mirror has a 3-degree field of view and is equipped with a CCD digital camera with 1.4 billion pixels. The regular observing of Pan-STARRS1 has already started in 2009. As one of science goals, Pan-STARRS1 studies dark energy through the SN, the WL and the CL techniques.

The South Pole Telescope (SPT) [8] is an international collaboration program operating at the USA NSF South Pole research station. As the largest telescope ever deployed at the South Pole, SPT is a 10-m submillimeter-wave telescope with a 1000-element bolometric focal plane array. Constructed between 2006 and 2007, it is conducting a survey of galaxy clusters over 4000 $\deg^2$ using the Sunyaev–Zel'dovich effect (SZE). About 20000 clusters are expected to be discovered by the SPT in recent years.

The Hubble Space Telescope (HST) [9] is one of the most famous telescopes in the world. It was built by the USA National Aeronautics and Space Administration (NASA), with contributions from the European Space Agency (ESA). HST has two mirrors, the primary mirror diameter is 2.4 m, and the secondary mirror diameter is 0.3 m. Since launched in 1990, many HST observations have led to great breakthroughs in astrophysics, such as the discovery of the currently observed cosmic acceleration [10, 11]. So far, utilizing the HST, several SNIa datasets have been obtained, such as Gold [12, 13], Union [14], Constitution [15] and Union2 [16]. The present SN survey project of HST, HST Cluster Supernova Survey [17], is targeting supernovae in high redshift galaxy clusters at $0.9 < z < 1.5$. It is expected that more than 100 high redshift SNIa will be found in recent years.

The Planck [18] is a CMB satellite, which is created as the third Medium-Sized Mission of ESA's Horizon 2000 Scientific Programme. The telescope is an off-axis, aplanatic design with two elliptical reflectors and a 1.5 m diameter. It is designed to image the CMB anisotropies and polarization maps over the whole sky, with unprecedented sensitivity and angular resolution. While the WMAP data reach 200 billion samples after its nine-year mission, just a single year of observing by Planck will yield 300 billion samples. Besides pinning down the cosmological parameters of dark energy, Planck will also detect thousands of galaxy clusters using the SZE. It was launched in 2009, and is expected to yield definitive data by 2012.

17.2   Intermediate-Scale, Near-Future Projects

In this section we introduce the intermediate-scale, near-future projects. A list of some most representative projects of Stage III is given in Table 17.2.

Table 17.2   Intermediate-Scale, Near-Future Projects

Survey	Location	Description	Probes
HETDEX	Davis Mountains (USA)	Optical, 9.2-m	BAO
DES	Cerro Pachon (Chile)	Optical, 4-m	BAO, SN, WL, CL
Pan-STARRS4	Hawaii (USA)	Optical, 1.8-m×4	SN, WL, CL
ALPACA	Cerro Pachon (Chile)	Optical, 8-m	SN
BigBOSS	Tucson and Cerro Pachon	Optical, 4-m	BAO

The Hobby–Eberly Telescope Dark Energy Experiment (HETDEX) [19] is an international collaboration program led by University of Texas, Austin. It will use the 9.2-m Hobby–Eberly Telescope at McDonald Observatory to measure BAO over two areas, each 100 deg^2, using one million galaxies over the redshift range $1.8 < z < 3.7$. It is }expected that the survey will begin in 2012 and last for 3 years.

The Dark Energy Survey (DES) [20] is an international collaboration program led by Fermilab. It is a new 570 megapixel digital camera (DE-Cam) mounted on the 4-m Blanco Telescope of the Cerro Tololo Inter-

American Observatory (CTIO) in Chile. The DES plans to obtain photometric redshifts in four bands, and the planned survey area is 5000 deg^2. It will probe dark energy by using the BAO, the CL, the SN, and the WL technologies. The proponents will begin in 2012 and take 5 years to complete.

The Pan-STARRS4 is the updated version of Pan-STARRS1 [7]. It is a large optical survey telescope to be sited on Mauna Kea in Hawaii. It consists of an array of four 1.8-m telescopes, each mirror will have a 3 degree field of view and be equipped with a CCD digital camera with 1.4 billion pixels. The dark energy science goals of Pan-STARRS4 include SN, WL and CL surveys. The survey will continue for ten years.

The Advanced Liquid-mirror Probe for Astrophysics, Cosmology and Asteroids (ALPACA) [21] is a proposed wide-field telescope employing an 8-meter rotating liquid mirror. It brings together the technologies of liquid-mirrors, lightweight conventional mirrors, and advanced detectors, to make a novel telescope with uniquely-powerful capabilities. Scanning a long strip of sky every night at CTIO, ALPACA will discover ~ 50000 SNIa and ~ 12000 type II SN each year to a redshift $z \sim 0.8$.

The Big Baryon Oscillation Spectroscopic Survey (BigBOSS) [22] is a USA NSF/DOE collaboration program. As the updated version of SDSS BOSS, BigBOSS will probe dark energy through the BAO and the redshift distortions techniques. It will build a new 4000-fiber spectrograph covering a 3-degree diameter field. This instrument will be mounted on the 4-meter Mayall telescope at Kitt Peak National Observatory (KPNO) for a 6-year run, then will be moved to the 4-m CTIO Blanco Telescope for another 4-year run. After 10-year operation, BigBOSS will complete the survey of 50 million galaxies and 1 million quasars from $0.2 < z < 3.5$ over 24000 deg^2. The operation of the survey will start in 2015.

17.3   Larger-Scale, Longer-Term Future Projects

In this section we introduce the larger-scale, longer-term future projects. A list of some most representative projects of Stage IV is given in Table 17.3.

A most ambitious ground-based dark energy survey project is the Large Synoptic Survey Telescope (LSST) [23], which is a USA NSF/DOE collaboration program. LSST is an 8.4-meter ground-based optical telescope (with a 9.6 deg^2 field of view and a 3.2 Gigapixel camera) to be sited in Cerro Pachon of Chile. Over a 10-year lifetime, it will obtain a database including

10 billion galaxies and a similar number of stars. As one of the most important scientific goals, LSST will study dark energy through a combination of the BAO, the SN and the WL techniques. Since its compelling scientific capacity and relatively low technical risk, LSST was selected as the top priority large-scale ground-based project for the next decade of astronomy in the Astro2010 report [24]. The appraised construction cost of LSST is 465 million U.S. dollars. The project is scheduled to have first light in 2016 and the beginning of survey operations in 2018.

Table 17.3   Larger-Scale, Longer-Term Future Projects

Survey	Location	Description	Probes
LSST	Cerro Pachon (Chile)	Optical, 8.4-m	BAO, SN, WL
SKA	Australia or South Africa	Radio, km^2	BAO, WL
WFIRST	Sun–Earth L2 Orbit	Infrared, 1.5-m	BAO, SN, WL
Euclid	Sun–Earth L2 Orbit	Optical/NIR, 1.2-m	BAO, WL
IXO	Sun–Earth L2 Orbit	X-ray	CL

Another ambitious large-scale ground-based telescope is the Square Kilometer Array (SKA) [25], which is a international collaboration program. SKA has a collecting area of order one square kilometer and a capable of operating at a wide frequency range (60 MHz–35 GHz). It will be built in the southern hemisphere (either in Australia or in South Africa), and the specific site will be determined in 2012. SKA will probe dark energy by BAO and WL techniques via the measurements of the Hydrogen line (HI) 21-cm emission in normal galaxies at high redshift. The total Budget of SKA is 1.5 billion U.S. dollars. Construction of the SKA is scheduled to begin in 2016 for initial observations by 2020 and full operation by 2024.

A most exciting space-based dark energy project is the Wide Field Infrared Survey Telescope (WFIRST) [26], which is a USA NASA/DOE collaboration program. WFIRST is a 1.5-meter wide-field near-infrared space telescope, orders of magnitude more sensitive than any previous project. The design of WFIRST is based on three separate inputs (JDEM-Omega, the Microlensing Planet Finder, and the The Near-Infrared Sky Surveyor) to Astro2010 [27]. It will enable a major step forward in dark energy understanding, provide a statistical census of exoplanets and obtain an ancillary data set of great value to the astronomical community. The objective of the WFIRST dark energy survey is to determine the nature of dark energy by measuring the expansion history and the growth rate of large scale structure. A combination of the BAO, the SN and the WL techniques will be used to probe dark energy. Since its compelling scientific capacity and relatively low technical risk, WFIRST was selected as the top priority large-scale ground-based project for the next decade of astronomy in the Astro2010 report [24]. The appraised construction cost of WFIRST is 1.6 billion U.S. dollars. The project is scheduled to launch in 2020 and has a 10-years lifetime.

Another exciting space-based dark energy project is the Euclid [28], which is an ESA project. Euclid is 1.2-m Korsch telescope, with optical and near-infrared (NIR) observational branch. The primary goal of Euclid is to map the geometry of the dark universe, and it will search galaxies and clusters of galaxies out to $z \sim 2$, in a wide extragalactic survey covering 20000 $\deg^2$, plus a deep survey covering an area of 40 $\deg^2$. The mission is optimized for two primary cosmological probes: BAO and WL. It will also make use of several secondary cosmological probes such as the ISW, CL and redshift space distortions to provide additional measurements of the cosmic geometry and structure growth. After 5 years' survey, Euclid will measure the DE EOS parameters w_0 and w_a to a precision of 2% and 10%, respectively, and will measure the growth factor exponent γ with a precision of 2%.

The last Stage IV project is the International X-ray Observatory (IXO)[29], which is a partnership among the NASA, ESA, and the Japan Aerospace Exploration Agency (JAXA). IXO is a powerful X-ray space telescope that features a single large X-ray mirror assembly and an extendable optical bench with a focal length of ~ 20 m. With more than an order of magnitude improvement in capabilities, IXO will study dark energy through the WL technique. Because of IXO's high scientific importance, it was selected as the fourth-priority large-scale ground-based project in the Astro2010 report [24]. The total Budget of IXO is 5.0 billion U.S. dollars. The project is scheduled to launch in 2021.

References

[1] http://lhc.web.cern.ch/lhc.
[2] A. Albrecht et al., arXiv:astro-ph/0609591.
[3] Y. Wang, Phys. Rev. D **77** (2008) 123525.
[4] J. Frieman, M. Turner, and D. Huterer, Ann. Rev. Astron. Astrophys **46** (2008) 385.
[5] http://www.sdss3.org/cosmology.php.
[6] http://www.lamost.org.

[7] http://pan-starrs.ifa.hawaii.edu/public/; N. Kaiser *et al.*, Proc. SPIE, **4836** (2002) 154.
[8] http://pole.uchicago.edu/; J. E. Ruhl *et al.*, Proc. SPIE Int. Soc. Opt. Eng. **5498** (2004) 11.
[9] http://hubblesite.org/.
[10] A. G. Riess *et al.*, AJ. **116** (1998) 1009.
[11] S. Perlmutter *et al.*, ApJ. **517** (1999) 565.
[12] A. G. Riess *et al.*, ApJ. **607** (2004) 665.
[13] A. G. Riess *et al.*, ApJ. **659** (2007) 98.
[14] M. Kowalski *et al.*, ApJ. **686** (2008) 749.
[15] M. Hicken *et al.*, ApJ. **700** (2009) 1097; M. Hicken *et al.*, ApJ. **700** (2009) 331.
[16] R. Amanullah *et al.*, ApJ. **716** (2010) 712.
[17] K. S. Dawson *et al.*, ApJ. **138** (2009) 1271; K. Barbary *et al.*, arXiv: 1010.5786.
[18] http://www.rssd.esa.int/index.php?project=PLANCK.
[19] http://hetdex.org/.
[20] http://www.darkenergysurvey.org/; T. Abbott *et al.*, AIP Conf. Proc. **842** (2006) 989.
[21] http://www.astro.ubc.ca/LMT/alpaca/index.html/.
[22] http://bigboss.lbl.gov/; D. J. Schlegel *et al.*, arXiv:0904.0468.
[23] http://www.lsst.org/lsst.
[24] http://sites.nationalacademies.org/bpa/BPA 049810.
[25] http://www.skatelescope.org/.
[26] http://wfirst.gsfc.nasa.gov/.
[27] http://wfirst.gsfc.nasa.gov/science/.
[28] http://sci.esa.int/euclid/.
[29] http://ixo.gsfc.nasa.gov/.

18
Observational Constraints on Specific Theoretical Models

In this chapter we will briefly review some research works concerning the cosmological constraints on the specific theoretical models. These models can be classified into three classes: (i) Models that modify the energy-momentum tensor on the R.H.S. of the Einstein equation, i.e. dark energy models. We will briefly introduce some numerical results of scalar field models, Chaplygin gas models, and holographic models. (ii) Models that modify the L.H.S. of the Einstein equation, i.e. modified gravity models. We will briefly introduce some numerical results of the DGP braneworld model, the $f(\mathcal{R})$ gravity, the Gauss–Bonnet gravity, the Brans–Dicke theory and the $f(\mathcal{T})$ gravity. (iii) Models that attempt to explain the apparent cosmic acceleration by assuring the inhomogeneities in the distribution of matter. We will briefly introduce some numerical works on the inhomogeneous Lemaître–Tolman–Bondi model and back-reaction model. Lastly, we will summarize and compare these theoretical models.

18.1   Scalar Field Models

As is well known, the most popular phenomenological models are models with rolling scalar fields ϕ [1, 2]. These models are the direct generalization of the cosmological constant and have been well-studied both theoretically and numerically. In Part II, we have introduced the theoretical studies on the scalar field models. In this section, we will discuss the scalar field models from the aspect of observations.

- Reconstructing the scalar field models from observations.

After the proposal of quintessence field as a candidate of dynamical dark energy [1, 2], a lot of numerical studies have been carried out to distinguish this model from the cosmological constant (see [3–9] for some early studies). A widely used approach is directly reconstructing dark energy from the observational quantities like $d_{\rm L}(z)$, $H(z)$, and so on. In the quintessence model, the Einstein equations can be rewritten as [4, 6, 10]

$$\frac{8\pi G}{3H_0^2}V(x) = \frac{H^2}{H_0^2} - \frac{x}{6H_0^2}\frac{dH^2}{dx} - \frac{1}{2}\Omega_{\rm m0}x^3\,, \tag{18.1}$$

$$\frac{8\pi G}{3H_0^2}\left(\frac{d\phi}{dx}\right)^2 = \frac{2}{3H_0^2 x}\frac{d\ln H}{dx} - \frac{\Omega_{\rm m0}x}{H^2}\,, \quad x \equiv 1 + z\,. \tag{18.2}$$

One can determine $\phi(z)$, and thus its inversion $z(\phi)$ by integrating Eq. (18.2). The Hubble parameter $H(z)$ together with its first derivative can be determined from observations, such as the measurements of the luminosity distance $d_{\rm L}(z)$ [4–7, 10]

$$H(z) = \left[\frac{d}{dz}\left(\frac{d_{\rm L}(z)}{1+z}\right)\right]^{-1}\,. \tag{18.3}$$

Substituting $H(z)$ and $z(\phi)$ into Eq. (18.1), one can reconstruct the potential $V(\phi)$. In 2000 Saini *et al.* [6] reconstructed $V(\phi)$ and $w_{\rm Q}(z)$ based on 54 SNIa given by Perlmutter *et al.* [11]. Their results are shown in Figure 18.1.

In the same way one can also reconstruct other physical quantities from observations. For instance, in [6], Saini *et al.* reconstructed the EOS $w_{\rm de}(z)$ using the relation

$$w_{\rm de}(z) = \frac{(2/3)(1+z)d\ln H/dz - 1}{1 - (H_0/H)^2\Omega_{\rm m0}(1+z)^3}\,. \tag{18.4}$$

They obtained the 1σ constraint

$$-1 \leqslant w_{\rm de} \leqslant -0.86\,, \qquad -1 \leqslant w_{\rm de} \leqslant -0.66\,, \tag{18.5}$$

corresponding to the value of $w_{\rm de}$ at $z = 0$ and $z = 0.83$, respectively. So the result allows the possible evolution of dark energy, while the cosmological constant with $w_{\rm de} = -1$ is still consistent with the data.

In addition to the above method, a more widely used approach is to reconstruct the quantities by a fitting ansatz which relies on a small number of free parameters [13, 14]. We will discuss this issue in Chapter 19.

Broadly speaking, by making use of the scalar fields one can reconstruct a dark energy component with any property. So the issue of the observational tests of scalar field models is somewhat similar as the observational

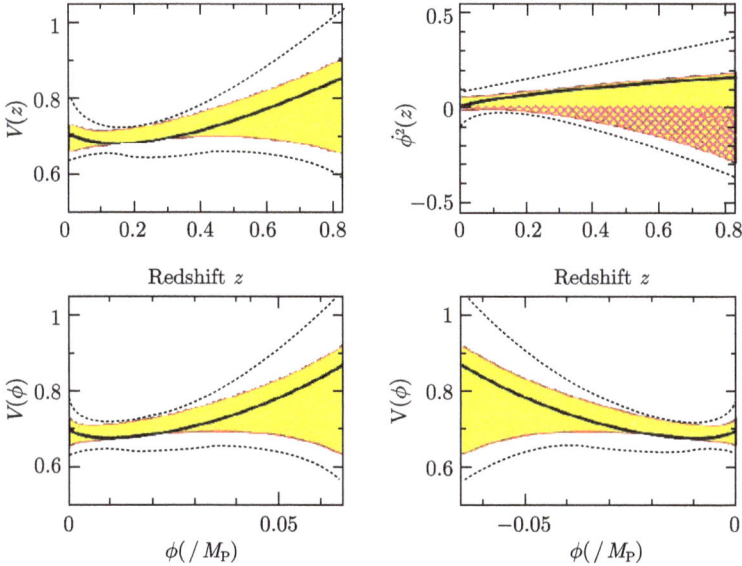

Figure 18.1   The reconstructed potential $V(z)$ and the kinetic energy term $\dot{\phi}^2$ in units of the critical density $\rho_{\rm cr} = 3M_{\rm P}^2 H_0^2$, based on 54 SNIa data of the SCP [11] including the low-z Calan Tololo sample [12]. Also plotted are the two forms of $V(\phi)$ for this $V(z)$. The value of ϕ, known up to an additive constant, is plotted in units of the Planck mass. The solid line corresponds to the best-fit values of the parameters, the shaded area/dotted lines covers the range of 68%/90% errors, respectively. The hatched area represents the unphysical region $\dot{\phi}^2 < 0$. From [6]. Reprinted figures with permission of Somak Raychaudhury. Copyright (2000) by the American Physical Society.

probe of the dynamical behavior of dark energy. One can see [15–20] and references therein for more studies on the scalar field models from the aspect of observations. In addition, there are also some theories using vector fields to describe dark energy [21–24].

• Quintessence, phantom or quintom.

Another interesting issue concerning the scalar field models is the future evolution of dark energy and the fate of our universe. Besides the quintessence field satisfying $w_{\rm de} > -1$, another well-studied scalar field model is the phantom model satisfying $w_{\rm de} < -1$. In this model, the dark energy density will reach infinity and lead to the "big rip" singularity. The WMAP7 measurements give $w_{\rm de} = -1.10 \pm 0.14$ [25], the analysis of the Union2 SNIa dataset gives $w_{\rm de} = -1.035^{+0.093}_{-0.097}$ [26], and the analysis of the SDSS DR7 gives $w_{de} = -0.97 \pm 0.10$ [27]. So the current observational data are still consistent with the cosmological constant, although the possibilities of quintessence and phantom all exist. Besides, the possibility of the quintom, where the EOS can cross $w_{de} = -1$, can also provide a consistent fit to the observations. We refer to [28] for related numerical studies.

• Interaction between dark sectors.

It is also worthwhile to consider the possible interaction between the scalar field and the matter component. In [29], Amendola proposed the coupled quintessence (CQ) model, in which the scalar field ϕ and the dark matter fluid with each other through a source term in their respective covariant conservation equations

$$\nabla_\mu T^\mu_{\nu(\phi)} = -Q_\nu, \qquad \nabla_\mu T^\mu_{\nu(m)} = Q_\nu, \qquad (18.6)$$

where $T^\mu_{\nu(\phi)}$ and $T^\mu_{\nu(m)}$ represent the stress tensors of ϕ and matter. In [29, 30], it was proposed that the source term Q_ν assumes the form

$$Q_\nu = -\kappa\beta(\phi)T_{(m)}\nabla_\nu\phi, \qquad (18.7)$$

where $T_{(m)}$ is the trace of $T^\mu_{\nu(m)}$, and $\beta(\phi)$ (hereafter β) is the coupling function that sets the strength of the interaction.

In [29], Amendola also investigated the evolution of the perturbations as well as the observational signature of this model. Hereafter, a lot of studies were performed to constrain the CQ model utilizing the various observational techniques [31–35]. For instance, in [32], from first-year WMAP observations, Amendola et al. obtained an upper constraint ~ 0.1 for the coupling parameter β. In [33], Maccio et al. performed the N-body simulations in CQ model and found that this model is consistent with the growth of structure in the non-linear regime. In [34], Guo et al. considered the following kinds of interaction

$$\dot\rho_m + 3H\rho_m = \delta H\rho_m. \qquad (18.8)$$

From a combination of SNIa, WMAP and BAO data, they put a stringent constraint on the constant coupling mode

$$-0.08 < \delta < 0.03, \qquad (18.9)$$

at the 2σ CL (see Figure 18.2). In [35], Bean et al. investigated a variety of CQ models and found that a combination of SNIa, LSS and CMB data

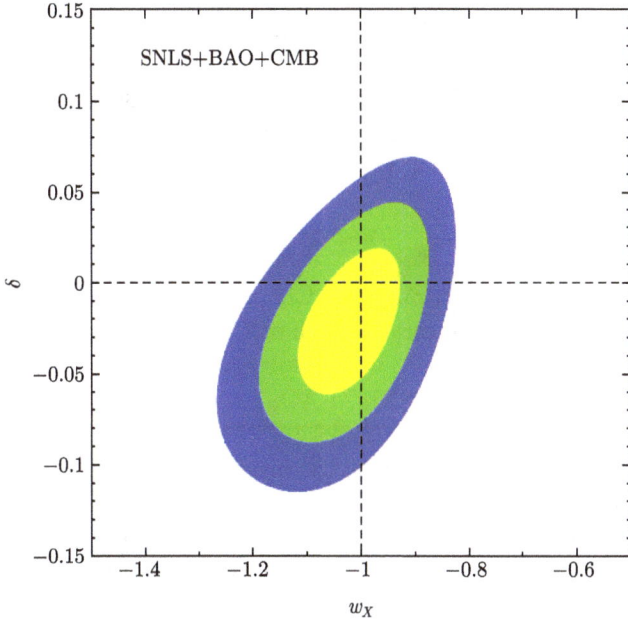

Figure 18.2   Observational constraints on the coupling between dark energy and dark matter using 71 SNIa from the fist year SNLS [36], the CMB shift parameter from the three-year WMAP observations [37], and the BAO peak found in the SDSS [38]. The constant coupling $\delta = \Gamma/H$ is considered, which leads to the Friedmann equation $\dot{\rho}_m + 3H\rho_m = \delta H\rho_m$. The left and right panels show observational contours in the (w_X, δ) and (Ω_{X0}, δ) planes, respectively (here "X" stands for the dark energy component). The best-fit model parameters correspond to $\delta = -0.03, w_X = -1.02$ and Ω_{X0}=0.73. A stringent constraint $-0.08 < \delta < 0.03$ at 95% CL is obtained. From [34]. Reprinted figure with kind permission of Shinji Tsujikawa and the APS. Copyright (2007) by the American Physical Society.

can constrain the strength of coupling between dark sectors to be less than 7% of the coupling to gravity. In all, the current observational data have already given tight constraints on the interaction between dark sectors. One can see [39] and references therein for more studies on this topic.

18.2 Chaplygin Gas Models

In [40], Kamenschchik *et al.* explained dark energy as a kind of fluid called Chaplygin gas (CG), characterized by the EOS

$$p_{de} = -\frac{A}{\rho_{de}}, \tag{18.10}$$

where A is a constant. Later, Bilic $et\ al.$ [41] and Bento $et\ al.$ [42] proposed an extension of the original Chaplygin gas model, called generalized Chaplygin gas (GCG) model, with the EOS

$$p_{de} = -\frac{A}{\rho_{de}^{\alpha}}. \tag{18.11}$$

• Observational inspection of the Chaplygin gas models.

The observational inspections of the CG and GCG models have been extensively investigated using various observational methods [43–46], including the observations of SNIa, BAO, CMB, WL, X-ray, GRB, X-ray, Fanaroff–Riley type IIb radio galaxies [47], and so on. A common result of these studies is that the CG model has been ruled out by the observations. For example, in [46], Davis $et\ al.$ showed that the CG model is strongly disfavored by a combination of SNIa+BAO+CMB data, since this model gives a $\chi^2_{min} \sim 100$ larger than that given by the ΛCDM model (see Figure 18.3 for more details). Similar result was obtained in [45], where Zhu investigated the GCG model from the measurements of X-ray, the SNIa and Fanaroff–Riley type IIb radio galaxies, and got a constraint

$$\alpha = -0.09^{+0.54}_{-0.33}, \tag{18.12}$$

at the 95% CL. Therefore, the CG model, which corresponds to $\alpha = 1$, is ruled out at a 99% CL.

• The GCG model as a unification of dark energy and dark matter.

An attractive feature of the GCG model is that it can explain both dark energy and dark matter in terms of a single component and has been referred to as "unified dark matter" (UDM) [48–50]. From Eq. (18.11) one can obtain the energy density of GCG [50]

$$\rho(t) = \left[A + \frac{B}{B^{3(1+\alpha)}} \right]^{\frac{1}{1+\alpha}}, \tag{18.13}$$

where B is an integration constant. Defining

$$\Omega^*_{m0} = \frac{B}{A+B}, \qquad \rho_* = (A+B)^{\frac{1}{1+\alpha}}, \tag{18.14}$$

then the Eq. (18.11) becomes

$$\rho(z) = \rho_0 \left[(1 - \Omega^*_{m0}) + \Omega^*_{m0} a^{-3(1+\alpha)} \right]^{\frac{1}{1+\alpha}}. \tag{18.15}$$

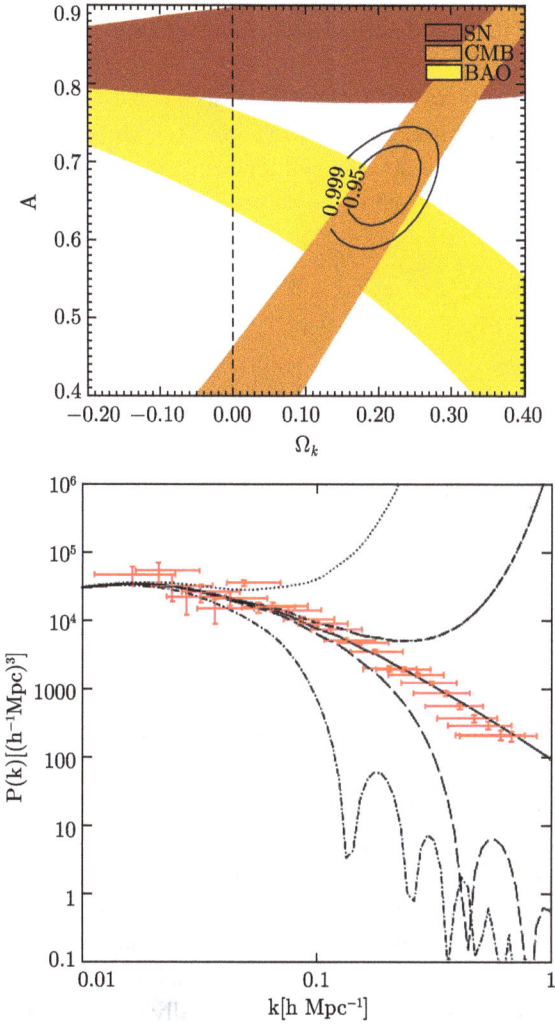

Figure 18.3   *Upper panel*: A constraint on the standard Chaplygin gas using SNIa, BAO and CMB. Clearly this model is a very poor fit to the data. From [46]. *Lower panel*: GCG as the unification of dark matter and dark energy will produce inconsistent oscillations in the mass power spectrum. The data points are the power spectrum of the 2df galaxy redshift survey, and the curves from top to bottom are GCG models with $\alpha = -10^{-4}$, -10^{-5}, 0 (ΛCDM), 10^{-5} and 10^{-4}, respectively. From [44]. Reproduced by permission of Tamara Davis and the AAS.

Obviously, the flat ΛCDM scenario with matter ratio Ω^*_{m0} is recovered with $\alpha = 0$.

However, a problem with the UDM scenario is that it will produce oscillations or exponential blowup of the matter power spectrum, which is inconsistent with observation. In [44], Sandvik *et al.* found that the density fluctuation δ_k with wave vector k evolves as

$$\delta''_k + [2 + \zeta - 3(2w - c_s^2)]\delta'_k = \left[\frac{3}{2}(1 - 6c_s^2 + 8w - 3w^2) - \left(\frac{kc_s}{aH}\right)^2\right]\delta_k\,,$$

(18.16)

where the quantity ζ, the EOS w and the squared sound speed c_s^2 take the form [44]

$$\zeta \equiv \frac{(H^2)'}{2H^2} = -\frac{3}{2}\left(1 + (1/\Omega^*_m - 1)a^{3(1+\alpha)}\right)^{-1}\,,$$

(18.17)

$$w = -\left[1 + \frac{\Omega^*_m}{1 - \Omega^*_m}a^{-3(1+\alpha)}\right]^{-1}\,,\qquad c_s^2 = -\alpha w\,,$$

(18.18)

and prime denotes partial differentiation with respect to $\ln a$. For Chaplygin gas, $\alpha > 0$ will result in $c_s^2 < 0$, which leads to a non-zero Jeans length $\lambda_J = \sqrt{\pi|c_s^2|/G\rho}$. In this regime the fluctuations will oscillate instead of grow polynomially. On the other hand, when $\alpha < 0$, perturbations will grow exponentially, which is also ruled out.

In [44], Sandvik *et al.* further confirmed the above arguments by numerical solving the equations (see Figure 18.3). By performing a χ^2 fit of the theoretically predicted power spectrum against that observed by the 2dF-GRS [51], they found that this inconsistency excludes most of the previously allowed parameter space of α, leaving essentially only the standard ΛCDM limit. This topic was also studied in [52, 53]. In [53], based on an analysis including the SNIa, the baryonic matter power spectrum, the CMB and the perturbation growth factor, Park *et al.* found that the allowed region for α is

$$-5 \times 10^{-5} \leqslant \alpha \leqslant 10^{-4}\,.$$

(18.19)

The allowed parameter space is extremely close to the ΛCDM model, so the possibility of the unification of dark matter and dark energy in the GCG scenario has been ruled out by the current cosmological observations.

18.3   Holographic Dark Energy Models

In the following, we will introduce some numerical works about the holographic dark energy (HDE) model, which arises from the holographic principle.

• The HDE model with the future event horizon as the cutoff.

The HDE has the form of energy density

$$\rho_{de} = 3c^2 M_P^2 L^{-2}, \tag{18.20}$$

where c is a number introduced in [54]. In 2004, Li [54] proposed to take $L = 1/R_h$, where R_h is the future event horizon defined as

$$R_h = a \int_a^\infty \frac{da}{Ha^2}. \tag{18.21}$$

That yields an EOS

$$w_{de} = -\frac{1}{3} - \frac{2}{3} \frac{\sqrt{\Omega_{de}}}{c}, \tag{18.22}$$

which satisfies $w_{de} < -1/3$ and can accelerate the cosmic expansion.

In [55], Huang and Gong first performed a numerical study on the HDE model. Making use of the 157 gold SNIa data, they obtained $\Omega_{m0} = 0.25^{+0.04}_{-0.03}$, $w_{de} = -0.91 \pm 0.01$ at 1σ CL for the $c = 1$ case. In [56], Zhang and Wu constrained the HDE model by performing a joint analysis of SNIa+CMB+LSS. They found that the best fit results are $c = 0.81$, $\Omega_{m0} = 0.28$, and $h = 0.65$, which implies that the HDE behaves as a quintom-type dark energy. In [57], Yi and Zhang tested the HDE model by using the $H(z)$ data and the LT data. They also verified that the HDE behaves like a quintom-type at 1σ CL. In addition, a lot of numerical studies were performed to test and constrain the HDE model [58]. These works showed that the HDE model can provide a good fit to the data. For example, in [59], by using the combined Constitution+BAO+CMB data, Li *et al.* obtained the following χ^2_{min}s for the ΛCDM and HDE models:

$$\chi^2_{\Lambda CDM} = 467.775, \qquad \chi^2_{HDE} = 465.912. \tag{18.23}$$

So the HDE model is consistent with the current observations. Similar results have been obtained in e.g. [60–63]. Therefore, from the perspective of current observations, HDE is a competitive model.

The parameter c plays an essential role in determining the evolution of the HDE. If $c = 1$, the dark energy EOS will be asymptotic to that of a

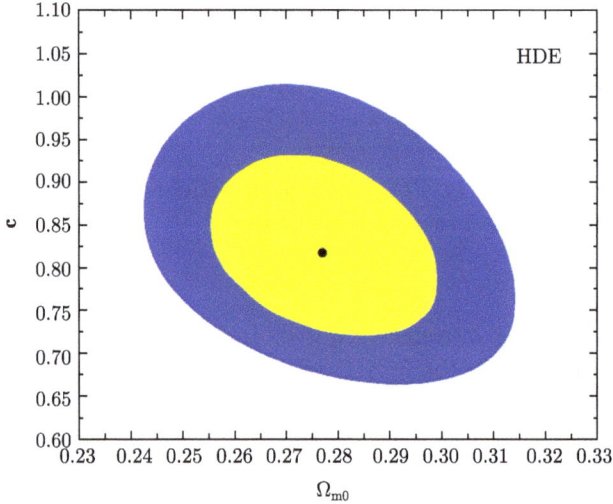

Figure 18.4   Constraints on the HDE model in Ω_{m0}-**c** plane from observational data including Constitution SNIa sample [64], the BAO data from SDSS [38], and the CMB data from WMAP5 measurements [37]. The data favor **c** < 1. For the constraint on **c**, the analysis gives $0.818^{+0.196}_{-0.154}$ at the 68.3% CL, and $0.818^{+0.196}_{-0.097}$ at the 95.4% CL. From [59].

cosmological constant and the universe will enter the de Sitter phase in the future; if **c** > 1, the EOS is always greater than −1, and HDE behaves as quintessence dark energy; if **c** < 1, initially the EOS of HDE is greater than −1, then it will decrease and eventually cross the $w_{de} = -1$ line, leading to a phantom universe with big rip as its ultimate fate. The numerical studies on HDE showed that the cosmological observations favors **c** < 1. For example, [59] gave **c** $= 0.818^{+0.113}_{-0.097}$ at the 68% CL (see Figure 18.4), while some later works [65, 63] with the improved data tighten the constraint to **c** < 0.9 at the 95% CL.

• Interactions between HDE and dark matter.

In [66, 67], Wang $et\ al.$ first studied the interaction between dark matter and the HDE. The introduction of interactions not only alleviates the cosmic coincidence problem,[1] but also avoids the future big rip singularity [61, 69–74]. Phenomenologically, the interaction can be introduced by

$$\dot{\rho}_m + 3H\rho_m = Q\,, \qquad \dot{\rho}_{de} + 3H(\rho_{de} + p_{de}) = -Q\,, \qquad (18.24)$$

---

[1] Coincidence problem may also be solved by the anthropic constraints [68].

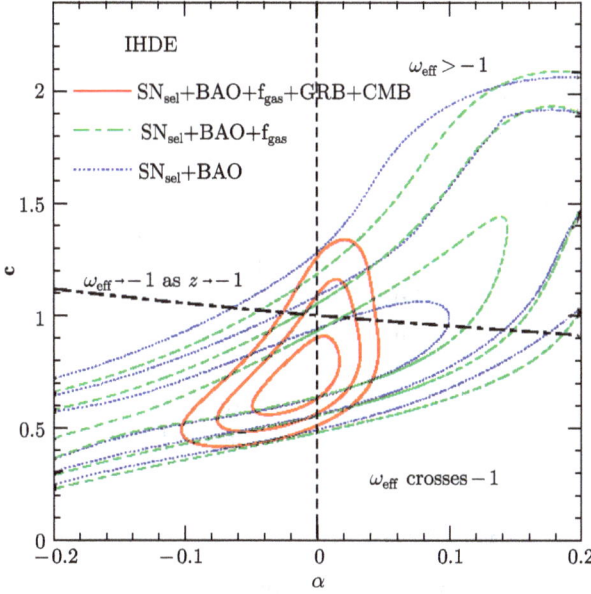

Figure 18.5 The contour maps of α vs. $\mathbf{c}$ for interacting HDE (IHDE) with 68%, 95.5% and 99.7% CL, obtained from a joint analysis including the Golden06+BAO+X-ray+GRB+CMB data. The black dot-dashed curve denotes $w_{\text{eff}} = -1$ when $z \to -1$ with $\Omega_{\text{de}0} = 0.73$, and the region below (over) it means w_{eff} will (not) cross -1 during infinite time. From [61], With kind permission of Yin-Zhe Ma and The European Physical Journal (EPJ).

where Q is the interaction term. Current observations have put tight constraints on the interactions. In [61], with the assumption of a flat universe, Ma considered the following Q and obtained the corresponding constraint

$$Q = 3\alpha H \rho_{\text{de}}, \qquad \alpha = -0.006^{+0.021}_{-0.024}, \qquad (18.25)$$

from a joint analysis (see Figure 18.5). In [73], Feng $et\ al.$ considered another class of Q and obtained the corresponding constraint

$$Q = 3bH(\rho_{\text{m}} + \rho_{\text{de}}), \qquad b = -0.003^{+0.012}_{-0.013}, \qquad (18.26)$$

from a combination of SNIa+BAO+CMB+Lookbacktime data. Later, in [74], Li $et\ al.$ revisited these two models and obtained

$$\alpha = (-6.1 \times 10^{-5})^{+0.025}_{-0.036}, \qquad b = (-1.6 \times 10^{-4})^{+0.008}_{-0.009} \qquad (18.27)$$

from a joint analysis of Constitution+BAO+CMB. They also found that there exists significant degeneracy between the phenomenological interaction and the spatial curvature in the HDE model.

- The ADE and RDE models.

In addition to the HDE model with future event horizon as the cutoff, the Agegraphic dark energy (ADE) model [75–77] and the Holographic Ricci dark energy (RDE) model [78] are also motivated by the holographic principle (the ADE model can also be obtained from the Károlyházy relation, see [75, 76] for details). In these two models, the IR cutoff length scale is given by the conformal time η and the average radius of the Ricci scalar curvature $|\mathcal{R}|^{-1/2}$, respectively. There have been some numerical studies on these two models [59, 79–83]. In general, these studies showed that the ADE and RDE models are not favored by current observations. For example, in [59], Li et $al.$ obtained

$$\chi^2_{\text{ADE}} = 481.694, \qquad \chi^2_{\text{RDE}} = 483.130. \tag{18.28}$$

The $\chi^2_{\min}$s of the ADE and RDE model are much larger than that of the ΛCDM and HDE model listed in Eq. (18.23), showing that these two models are not favored by observations. The results have been further confirmed in some later works [65, 63].

18.4   Dvali–Gabadadze–Porrati Model

The Dvali–Gabadadze–Porrati (DGP) braneworld model is a theory where gravity is altered at immense distances by the slow leakage of gravity off our three-dimensional universe [84]. In this model, the Friedmann equation is modified as

$$H^2 - \frac{H}{r_c} = \frac{8\pi G}{3} \rho_m, \tag{18.29}$$

where $r_c = (H_0(1 - \Omega_{m0}))^{-1}$ is the length scale beyond which gravity leaks out into the bulk. At early times, $Hr_c \gg 1$, the Friedmann equation of general relativity is recovered. In the future, $H \to 1/r_c$, the expansion is asymptotically de Sitter.

There is also a generalized phenomenological DGP model characterized by the Friedmann equation [85]

$$H^2 - \frac{H^\alpha}{r_c^{2-\alpha}} = \frac{8\pi G}{3} \rho_m, \tag{18.30}$$

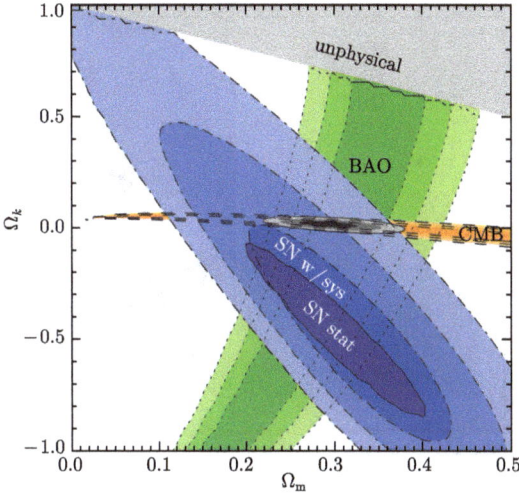

Figure 18.6 Combined SNIa+BAO+CMB observational constraint shows that DGP does not achieve an acceptable fit. The areas of intersection of any pair are distinct from other pairs. This is a strong signal that the DGP model is incompatible with the observations. From [86]. Reproduced by permission of Eric Linder and the AAS.

where $r_c = H_0^{-1}/(1 - \Omega_{m0})^{\alpha-2}$. This model interpolates between the pure ΛCDM model and the DGP model with an additional parameter α. $\alpha = 1$ corresponds to the DGP model and $\alpha = 0$ corresponds to the ΛCDM model.

• The DGP model is disfavored by the observations.

Although DGP is an attractive model allowing a self acceleration, many research works show that it is disfavored by observations [86–91]. For examples, in [86], from a joint analysis of SNIa+BAO+CMB, Rubin $et\ al.$ found that the DGP was disfavored by the data, with a $\Delta\chi^2=15$ compared with the ΛCDM model (see Figure 18.6). In [88], from a combination of SNIa and BAO measurements, Guo $et\ al.$ provided the constraints to the model parameters

$$\Omega_{m0} = 0.27^{+0.018}_{-0.017}, \quad \Omega_{r_c} = 0.216^{+0.012}_{-0.013}, \quad \Omega_{k0} = -0.350^{+0.080}_{-0.083} \quad (18.31)$$

at 99.73% CL. This result is in contradiction to the WMAP results indicating a flat universe. Moreover, the constraints to the generalized DGP model give a small α, indicating that the DGP is incompatible with the observations. In a recent work [92], Xia performed a joint analysis including SNIa, BAO, CMB, GRB and the linear growth factor of matter perturbations,

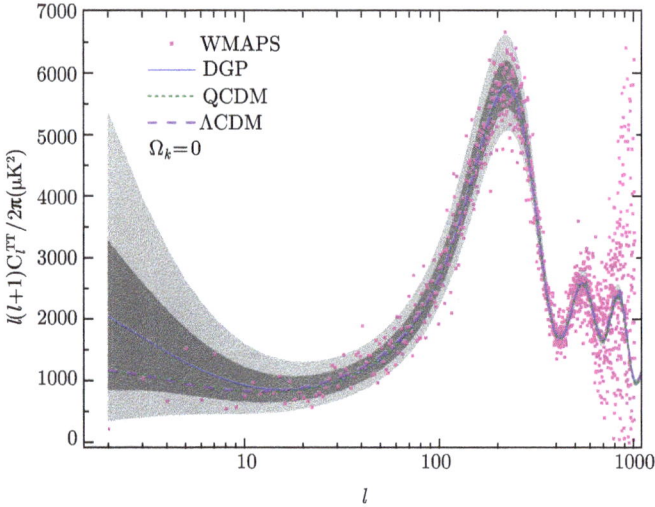

Figure 18.7   Predictions for the power spectra of the CMB temperature anisotropies C_l^{TT} of the best-fit DGP model (solid), a quintessence model with the same expansion history as DGP (short-dashed), and the ΛCDM model (dashed, coincident with the quintessence model at low l). Obtained by fitting to SNLS+ WMAP5+HST assuming a flat universe. Bands represent the 68% and 95% cosmic variance regions for the DGP model. Points represent WMAP5 measurements. It is clear that the best-fit DGP model over predicts the low-l modes anisotropy. From [87]. Reprinted figure with kind permission of Wenjuan Fang and the APS. Copyright (2008) by the American Physical Society.

and found a constraint

$$\alpha = 0.254 \pm 0.153 \qquad (18.32)$$

at the 68% CL, manifesting that this model tends to collapse to the cosmological constant when confronted with current observations.

• Testing the DGP model from the growth of structure.

As a modified gravity scenario, the growth of structure in the DGP gravity differs from that in the ΛCDM scenario. This can be used to test the DGP model [93, 94]. The perturbation theory in the DGP model has been studied [95–98]. These studies showed that the DGP gravity is disfavored by the observational data. For examples, in [96], Song et al. showed that the constraints from SNIa+CMB+H_0 exclude the simplest flat DGP model at about 3σ. Even including spatial curvature, best-fit open DGP model is a marginally poorer fit to the data than the flat ΛCDM model. In [87], Fang *et al.* showed that the DGP model is excluded at 4.9σ and 5.8σ levels with and without curvature respectively (see Figure 18.7). The corresponding $\chi^2_{\min}$s for the DGP and ΛCDM model are

$$\chi^2_{\Lambda\mathrm{CDM}} = 2777.8\,, \qquad \chi^2_{\mathrm{DGP}} = 2805.6\,. \tag{18.33}$$

The result is mainly due to the earlier beginning of the acceleration and the additional suppression of growth in the DGP scenario. In [99], by performing cosmological N-body simulations of the DGP model, Schmidt found that, independently of CMB constraints, the self-accelerating DGP model is strongly constrained by WL and cluster abundance measurements. Compared with the ΛCDM model, the abundance of halos above $10^{14}M_\odot$ is suppressed by more than a half in the DGP model. In all, when confronted with experiments, a lot of problems will emerge in the DGP model, indicating that this theory is strongly disfavored by the observations.

18.5 $f(\mathcal{R})$ Models

$f(\mathcal{R})$ gravity is a simplest modification to the general relativity with the replacement $\mathcal{R} \to f(\mathcal{R})$. In this section, we will focus on the observational tests of this theory. One can see [100–105] and references therein for more details.

- $f(\mathcal{R})$ gravity and its viable conditions.

At the beginning, it was proposed to explain the cosmic acceleration using the model with $f(\mathcal{R}) = \mathcal{R} - \alpha/\mathcal{R}^n (\alpha > 0, n > 0)$ [106–108]. However, later on, a lot of problems emerged in this model. In [109], it was shown that this model will lead to the matter instability. In [110], it was also found that this model is unable to satisfy local gravity constraints and pass the solar system tests of gravity. Much attentions have been paid to the analysis of the viable conditions of $f(\mathcal{R})$ models, and a lot of valuable results have been obtained [111–116]. For examples, to have a stable perturbation, the condition $f_{,\mathcal{R}\mathcal{R}} \equiv \partial^2 f/\partial\mathcal{R}^2 < 0$ is required; to have a stable late-time de Sitter point, the condition $0 < \mathcal{R}f_{,\mathcal{R}\mathcal{R}}/f_{,\mathcal{R}} < 1$ is also necessary. In summary, the conditions for the viability of $f(\mathcal{R})$ dark energy models include [100, 115, 116]:

(1) $f_{,\mathcal{R}} > 0$ and $f_{,\mathcal{R}\mathcal{R}} > 0$ for $\mathcal{R} \leqslant R_0$; \hfill (18.34)

(2) $f(\mathcal{R}) \to \mathcal{R} - 2\Lambda$ for $\mathcal{R} \ll \mathcal{R}_0$; \hfill (18.35)

(3) $0 < Rf_{,\mathcal{R}}/f_{,\mathcal{R}} < 1$ at the de Sitter point satisfying

$$\mathcal{R}f_{,\mathcal{R}} = 2f\,. \tag{18.36}$$

To be acceptable, an $f(\mathcal{R})$ model must satisfy the following conditions. Some viable models satisfying all these requirements have been proposed, and one can refer to e.g. [100, 103, 115–119] for more details about the viable conditions of $f(\mathcal{R})$ gravity.

- Cosmological tests of the $f(\mathcal{R})$ gravity.

We have listed some general conditions required for an $f(\mathcal{R})$ model to be valid. An $f(\mathcal{R})$ model satisfying these basic requirements, furthermore, should be confronted with the cosmological observations. A natural method is to test the $f(\mathcal{R})$ theories from the observations about the growth of structure, which depends on the theory of gravity. This issue attracted a lot of interests [113, 120–122]. Some observational signatures of the $f(\mathcal{R})$ models have been presented. For example, in [120, 121], Song *et al.* studied an $f(\mathcal{R})$ model parameterized by "Compton wavelength parameter" B, which is proportional to the second order derivative $f_{,\mathcal{R}\mathcal{R}}$,

$$B = \frac{f_{,\mathcal{R}\mathcal{R}}}{f_{,\mathcal{R}}} \frac{H d\mathcal{R}/d\ln a}{dH/d\ln a} .$$
(18.37)

The $B < 0$ branch violates the third condition of Eq. (18.34). For the stable $B > 0$ branch, it was found that this model will predict a lower large-angle CMB anisotropy by reducing the ISW effect, qualitatively change the correlations between the CMB and galaxy surveys, and alter the shape of the linear matter power spectrum (see Figure 18.8 for details).

To test the modified gravity theory, it is worthwhile to include various observational techniques probing gravity in different scales [123]. For example, in [124], Schmidt *et al.* conducted the first, simulation calibrated, cluster abundance constraints on a two-parameter modified action model,

$$f(\mathcal{R}) = \mathcal{R} - 2a\frac{\mathcal{R}}{\mathcal{R} + \mu^2} .$$
(18.38)

They found that the local cluster abundance, when combined with the data from CMB, SNIa, H_0 and BAO, can lead to a very tight constraint to the model, improving the previous constraints by $3 \sim 4$ orders of magnitude. The reason is that, the inclusion of cluster abundance data improves the bounds on the range of force modification from the several Gpc scale to tens of Mpc scale. In [125], Lombriser *et al.* revisited the model studied in [120, 121] and reported a strong constraint to the current value of the Compton wavelength

$$B_0 < 1.1 \times 10^{-3}$$
(18.39)

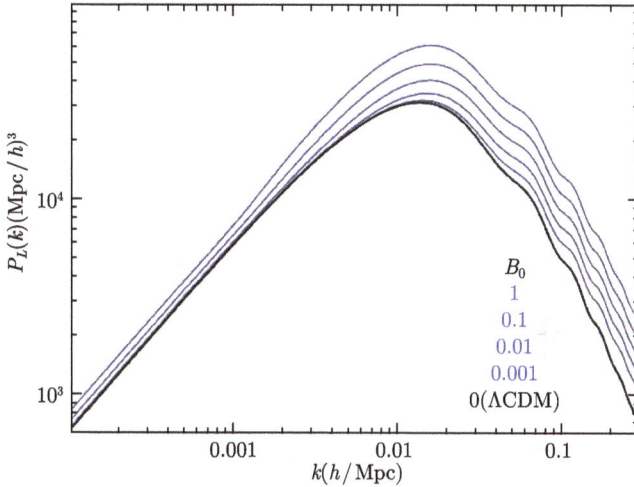

Figure 18.8   Linear matter power spectrum for several values of B_0 in the ΛCDM expansion history in the model [120, 121]. $\sqrt{B}$ is of order the Compton wavelength of the new gravitational degree of freedom, measured in the unit of Hubble length. For length scale larger than $\sqrt{B}/H$, the modified gravity effect is suppressed because of the mass of the scalar graviton. For length scales smaller than $\sqrt{B}/H$, gravity is modified. The modification of gravity leads to an enhancement in the growth of perturbations and a corresponding suppression for the decay of the gravitational potential. From [121]. Reprinted figure with kind permission of Wayne Hu and the APS. Copyright (2007) by the American Physical Society.

at the 2σ CL (see Figure 18.9), mainly due to the inclusion of data from cluster abundance.

Some other observational techniques have also been used to study the $f(\mathcal{R})$ models, such as the cosmic shear experiments [126, 127], the "21 cm intensity mapping" (detection of LSS in three dimensions without the detection of individual galaxies) [128, 129], the variation of the fine structure "constant" [130, 131], and so on. Nowadays, the observational tests of $f(\mathcal{R})$ gravity has drawn increasingly attention. For more details on the test of modify gravity theories from observations, see [132, 123, 133] and the references therein.

• The curvature singularity problem in $f(\mathcal{R})$ gravity. In [134, 135], it was proposed that there is curvature singularity problem in $f(\mathcal{R})$ theories, due to the dynamics of the effective scalar degree of freedom in the strong gravity regime. This problem leads to the contradiction with the existence

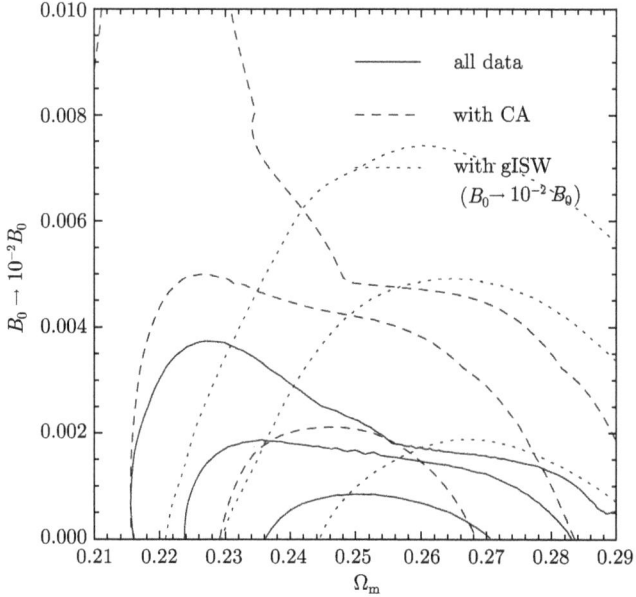

Figure 18.9   Contours of 2D marginalized 68%, 95% and 99% confidence bound-
aries using observational data from the measurements of CMB, BAO, Hubble
constant, and so on. For the constraints from galaxy-ISW (gISW) cross correlation, the Compton wavelength is rescaled as $B_0 \to 10^{-2} B_0$. From [125].
Reprinted figure with kind permission of Lucas Lombriser.

of the relativistic stars (like neutron stars) [136]. Furthermore, in [137], it
was shown that this problem can be cured via the addition of $\mathcal{R}^2$ term.
One can see Ref. [138] for some studies on this topic.

• $f(\mathcal{R})$ theories with Palatini approach.

The observational constraints on $f(\mathcal{R})$ theories within Palatini approach
were firstly performed by Amarzguioui [139] et al., where they investigated
the parameterization with the form of $f(\mathcal{R}) = \mathcal{R} + \alpha \mathcal{R}^\beta$ using the cosmological measurements SNIa+BAO+CMB. The best fit values are found to
be $\beta = 0.09$ with the allowed values rage

$$|\beta| < 0.2 \qquad (18.40)$$

at 1σ CL, and the previously commonly considered $1/\mathcal{R}$ model corresponding to $\beta = -1$ is ruled out by the constraint. In [140], using the same form
of parameterization, Koivisto calculated the power spectrum in the Palatini formulation of $f(\mathcal{R})$ gravity. By comparing the results to the SDSS

data [141], it was found that the observational constraints reduce the allowed parameter space to

$$|\beta| < \sim 10^{-4}, \tag{18.41}$$

which is a tiny region around the ΛCDM. Besides, the Palatini $f(\mathcal{R})$ gravity is also faced with some problems associated with non-dynamical nature of the scalar-field degree of freedom. One can refer to [142, 143, 144] for more studies about the Palatini $f(\mathcal{R})$ gravity.

18.6   Other Modified Gravity Models

We have introduced the observational aspects of the DGP scenario and the $f(\mathcal{R})$ models. In this section, we will discuss some other modified gravity theories, including the Gauss–Bonnet gravity, the Brans–Dicke gravity and the $f(\mathcal{T})$ gravity.

• Gauss–Bonnet gravity.

When compared with observations, many problems arise in the Gauss–Bonnet model [145]. In [146, 147], from the data analysis including the constraints from BBN, LSS, BAO and solar system data, Koivisto and Mota found that this model is strongly disfavored by the observations. In [148] and [149], Li *et al.* and De Felice *et al.* investigated the Gauss–Bonnet gravity, and found that the growth of perturbations gets stronger on smaller scales. This is incompatible with the observed galaxy spectrum, unless the deviation from the Einstein gravity is very small. Thus, the Gauss–Bonnet models have been effectively ruled out.

• Brans–Dicke theory.

Another well-known modified gravity model is the Brans–Dicke theory [150]. For the Brans–Dicke theory, the PPN parameter γ (see Eq. (16.43)) takes the form

$$\gamma = (1 + \omega)/(2 + \omega), \tag{18.42}$$

where ω is the constant in Eq. (9.79). From solar system and binary pulsar observations, the parameter ω has been tightly constrained to [151]

$$\omega \geqslant 4 \times 10^4, \tag{18.43}$$

and the cosmological effects of the scalar field are rendered insignificant. Later on, the Brans–Dicke theory is generalized to the scalar tensor theory [152]. One can see [100, 123, 103, 153, 154] and the references therein for more details on the test of this model.

- $f(\mathcal{T})$ theory.

There have been some numerical studies on the recently proposed $f(\mathcal{T})$ gravity [155, 156] as an explanation of the cosmic acceleration. In [157], Wu and Yu examined the following models from the Union2 SNIa dataset together with the BAO and CMB data,

$$f(\mathcal{T}) = \alpha(-\mathcal{T})^n \,, \qquad f(\mathcal{T}) = -\alpha\mathcal{T}(1 - e^{m\mathcal{T}_0/\mathcal{T}}) \,, \qquad (18.44)$$

and obtained the constraint $n = 0.04^{+0.22}_{-0.33}$, $m = -0.02^{+0.31}_{-0.20}$ at the 95% CL. They also compared the two models with the ΛCDM by using the $\chi^2_{min}/$dof (dof: degree of freedom) criterion. The results showed that ΛCDM is mildly favored by the data. Later, the power-law model was revisited by Bengochea [158], with the inclusion of GRB and H_0 data into consideration. In [159], Wei, Ma, and Qi tried to constrain $f(\mathcal{T})$ theory by using the varying fine structure "constant", and found that the observational $\Delta\alpha/\alpha$ data make $f(\mathcal{T})$ theory almost indistinguishable from the ΛCDM model. In addition, in [160], they also constrained $f(\mathcal{T})$ theory by using the varying gravitational "constant". It is found that the allowed model parameter n has been significantly shrunk to a very narrow range around 0.

In [161], Bamba *et al.* studied the cosmological evolution of the EOS for dark energy w_{de} in the exponential and logarithmic as well as their combination $f(\mathcal{T})$ theories. They found that the crossing of the phantom divide line of $w_{de} = -1$ can be realized in the combined $f(\mathcal{T})$ theory even though it cannot be in the exponential or logarithmic $f(\mathcal{T})$ theory. Moreover, the crossing is from $w_{de} > -1$ to $w_{de} < -1$, which is favored by the recent observational data.

The perturbations in $f(\mathcal{T})$ gravity has been studied in [162–164]. In [164], Zheng and Huang derived the evolution equation of growth factor for matter over-dense perturbation in $f(\mathcal{T})$ gravity. In addition, a problem in $f(\mathcal{T})$ gravity was pointed out by Sotiriou *et al.* in [165], where they showed that the Lorentz symmetry can not be restored in $f(\mathcal{T})$ theories due to sensible dynamics.

18.7   Inhomogeneous LTB and Back-Reaction Models

The inhomogeneous models have gathered significant interest in recent years as a scenario to explain the cosmological observations without invoking dark energy. In this section, we will briefly discuss the observational signature of the LTB model and the back-reaction model.

- LTB models.

The LTB models [166, 167] are the most commonly considered inhomogeneous scenario to explain the cosmic acceleration without introducing the dark energy component. For LTB models the structure of our universe is described by the inhomogeneous isotropic Lemaître–Tolman–Bondi (LTB) metric [168–170]

$$ds^2 = -dt^2 + \frac{R'(r,t)^2}{1+\beta(r)}dr^2 + R^2(r,t)(d\theta^2 + \sin^2\theta d\phi^2) \,, \qquad (18.45)$$

where the prime denotes partial differentiation with respect to r. $\beta(r)$ is a function of r. Notice that the FRW metric is recovered by requiring $R = a(t)r$ and $\beta = -kr^2$. The Hubble parameters at the transverse and radial direction are expressed as

$$H_\perp = \frac{\dot{R}'}{R'} \,, \qquad H_{/\!/} = \frac{\dot{R}}{R} \,, \qquad (18.46)$$

and the apparent cosmic acceleration can be explained by choosing suitable form of $R(r,t)$, without the introduction of dark energy [171]. In the simplest class of such models we live close to the center of a huge, spherically symmetric Gpc scale void. Due to the spatial gradients in the metric, our local region has a larger Hubble parameter than the outer region [172]. In some more complicated scheme, it has been proposed to reconstruct the cosmological constant in an inhomogeneous universe [173]. The idea is that, since cosmological observations are limited on the light cone, it is possible to reconstruct an inhomogeneous cosmological model (indistinguishable from the homogeneous ΛCDM model) to explain the cosmic acceleration without a cosmological constant. So far there have been some studies on the observational tests of the LTB model. In the following, we will briefly review some related works.

It has been shown that the void model is able to provide a good fit to the SNIa data. For example, in [174], Sollerman *et al.* tested two kinds of LTB models and compared them with the ΛCDM model. The models they considered have the following matter distribution,

$$\Omega_{\mathrm{m}}(r) = \Omega_{\mathrm{out}} + (\Omega_{\mathrm{in}} - \Omega_{\mathrm{out}})e^{-(r/r_0)^2} \,, \qquad (18.47)$$

$$\Omega_{\mathrm{m}}(r) = \Omega_{\mathrm{out}} + (\Omega_{\mathrm{in}} - \Omega_{\mathrm{out}})\left(\frac{1+e^{-r_0/\Delta_r}}{1+e^{(r-r_0)/\Delta_r}}\right) \,, \qquad (18.48)$$

where Ω_{in} is the matter density at the center of the void, Ω_{out} is the asymptotic value of the matter density outside the void, and r_0 is the size the underdensity. The second model [175] has a much sharper transition of matter density than the first one, with the extra parameter Δ_r characterizing the transition width. Using the first-year SDSS-II SNIa data [176] together with the BAO and CMB measurements, they found that χ^2 values for the LTB fits are comparable to that of the ΛCDM model,

$$\chi^2_{\Lambda CDM} = 233.2\,, \quad \chi^2_{LTB\ model1} = 235.5\,, \quad \chi^2_{LTB\ model2} = 237.6 \quad (18.49)$$

while the extra parameters in the LTB models make the models fare poorly in the information criteria tests.

However, a lot of problems arise in the LTB model when confronted with some other cosmological observations. In [177], Alnes et al. pointed out that a problem of the void model is that it requires us to live precisely near the center of the void. If there is a deviation between our position and the center of the void, the observed CMB dipole would become much larger than that allowed by observations (see the upper panel of Figure 18.10 for a brief description). Currently, the maximum distance to the center has been constrained to be very small [178–180],

$$r_{obs} <\sim 20 \ \ Mpc\,, \quad (18.50)$$

which leads to a fine-tuning problem. For example, in [181] Blomqvist and Mörtsell tested the LTB described by Eq. (18.47) by using the SNIa and CMB dipole data. They found that the position of the observer has been confined to within about one percent of the void scale radius (see the lower panel of Figure 18.10).

Besides, even if we happen to live very close to the center of the void, there should be some off-centered galaxy clusters where a large CMB dipole can be observed in their reference frame. For us, the relative motion between the CMB frame and the matter frame manifests itself observationally as a kinematic Sunyaev–Zel'dovich effect. In [182], Bellido et al. demonstrated that the limited observations of only 9 clusters with large error bars already rule out LTB models with void sizes greater than ~ 1.5 Gpc and a significant underdensity.

Another problem with the void models is that when fitted to the data, they slow local expansion values [179, 180, 183, 184, 185]

$$h \sim 0.45\text{--}0.6\,, \quad (18.51)$$

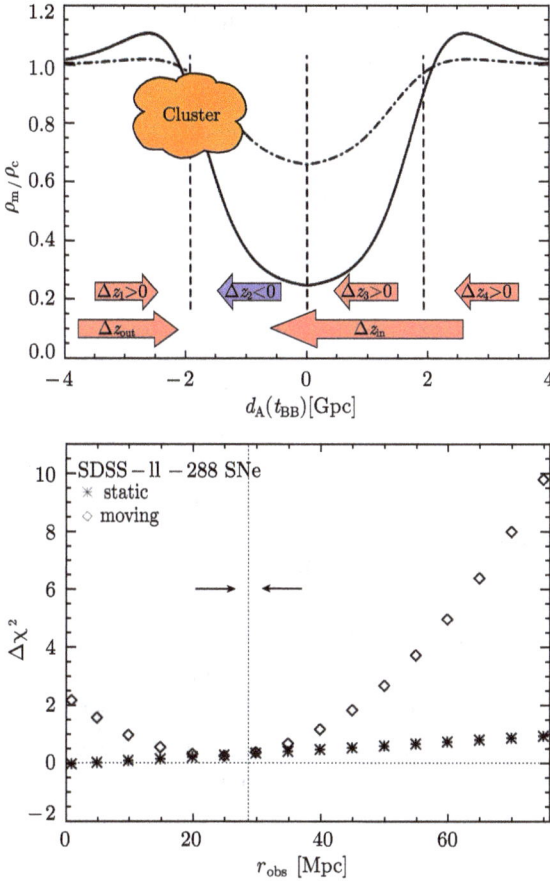

Figure 18.10 *Upper panel*: An off-center galaxy cluster in a void will observe a dipole in the CMB. Since the expansion rate inside the void is higher, photons arriving through the void (from the right in the figure) will have a larger redshift (Δz_{in}) than photons did not pass through the void (left, with redshift Δz_{out}). From [182]. Reprinted with kind permission of Troel Haugbølle and IOP Publishing Ltd. Copyright (2008) IOP Publishing Ltd and Sissa Medialab srl. *Lower panel*: The changes in the χ^2 values as a function of the observer's position. The SDSS-II data [176] combined with the CMB dipole requirement is used. The χ^2 value quickly increases as the observer is displaced away from the center. See [181] for details about the meaning of the diamonds, stars, lines and arrows. From [181]. Reprinted with kind permission of Edvard Mortsell and IOP Publishing Ltd. Copyright (2008) IOP Publishing Ltd and Sissa Medialab srl.

which seems to be in contradiction with the measurements of the local Hubble constant. Recently, in [186], Riess *et al.* obtained

$$H_0 = 73.8 \pm 2.4 \ (\mathrm{km/s})/\mathrm{Mpc}\,, \qquad (18.52)$$

corresponding to a 3.3 % uncertainty. They found that the void models with $h_0 \sim 0.6$ is ruled out by the measurements in more than 5σ.

There have been some other useful methods proposed to distinguish the void models between other inhomogeneous dark energy models. In [187], Uzan *et al.* presented the redshift drift $\dot{z}$ in a general spherically symmetric spacetime, and demonstrated that its observation would allow the test of Copernican principle. In addition, Yoo [188] and Quartin [189] showed that the dz/dt in void modes is always negative, which is greatly different from other dark energy models. Another interesting idea is, the ionized universe severs as a mirror to reflect CMB photons in other regions of the universe to us, and thus can tell us deviation from the Copernican principle [190, 191]. Utilizing this method, some models with largest voids have been excluded [191]. There are also some other methods of testing the LTB model, such as the scalar perturbations [184], the slope of low-z SNIa distance moduli [192], the constant curvature condition [193], the small scale CMB [194], the cosmic neutrino background [195], the cosmic age test [196], and so on.

In all, in recent years the research of the LTB model has drawn a lot of interests. Although this model can provide a good explanation of the SNIa data without introducing the mysterious dark energy component, when confronted with the CMB, H_0 and some other cosmological observations, a lot of problems emerged. For more details about the LTB model see [100, 196–198] and references therein.

• Back-reaction model.

Like the LTB model, the back-reaction model [199–203] is another model in which the cosmic expansion is due to the effect of the inhomogeneities of the universe. There have been some works concerning the possible observational signature of this model [204–206]. For example, in [204], Li *et al.* showed a nontrivial scale dependence of the Hubble rate in this model. In [206], Larena *et al.* proposed to use a template metric to deal with observations in back-reaction context and found that averaged inhomogeneous models can reproduce the observations of SNIa data and the position of the CMB peaks. To provide evidence for the back-reaction mechanism, further studies are needed. One can see [197, 207] for mode studies concerning the back-reaction model.

18.8   Comparison of Dark Energy Models

Facing so many dark energy candidates, it is very important to decide which one is more favored by the observational data. So far there have been many works on the comparison of various dark energy models [65, 46, 63, 86, 208–211]. We will briefly review the topic of model comparison in this section.

• Model selection and the information criteria.

The χ^2 statistics alone cannot provide an effective way to make a comparison between competing dark energy models. To do this, we should take into account the relative complexity of the models. To give a blatant example, a 10th-order polynomial will always give an equal or better fit than a straight line to any data set, but this does not mean that any of the extra eight coefficients have any significance. It just means that a model with more parameters will generally give an improved fit (always, if the simpler model is a subclass of the more complex one) [46].

To enforce a model comparison, a general way is to employ the information criteria (IC) to assess different models [212–214]. These statistics favor models that give a good fit with fewer parameters. The most frequently used IC including the Bayesian information criterion (BIC) [212] and the Akaike information criterion (AIC) (18.54). They are defined as

$$\mathrm{BIC} = -2\ln \mathbf{L}_{\mathrm{max}} + n_{\mathrm{p}} \ln N \,, \tag{18.53}$$

$$\mathrm{AIC} = -2\ln \mathbf{L}_{\mathrm{max}} + 2k \,, \tag{18.54}$$

where $\mathbf{L}_{\mathrm{max}}$ is the maximum likelihood (under Gaussian assumption $\chi^2_{\mathrm{min}} = -2\ln \mathbf{L}_{\mathrm{max}}$), n_{p} is the number of parameters, and N is the number of data points used in the fit. According to these criteria, models that give a good fit with fewer parameters will be more favored. So these criteria embody the principle of Occam's razor, "entities must not be multiplied beyond necessity". Generally, a ΔBIC of more than 2 (or 6) is considered positive (or strong) evidence against a model [215]. It should be noted that the IC alone can at most say that a more complex model is not necessary to explain current data, since a poor information criterion result might arise from the fact that the current data are too limited to constrain the extra parameters in this complex model, and it might become preferred with improved data.

In addition to the AIC and BIC, a more sophisticated method for model selection is the so-called Bayesian evidence (BE), which considers the increase of the allowed volume in the data space due to the addition of extra parameters rather than simply counting parameters. So it requires an integral of the likelihood over the whole model parameter space,

$$BE = \int \mathcal{L}(\mathbf{d}|\mathbf{p}, M)P(\mathbf{p}|M)d\mathbf{p}.$$ (18.55)

In practice, the integral in Eq. (18.55) can be calculated using the nested sampling method [216]. BE has already been applied in cosmology [217, 218]. Since in most cases the simpler AIC and BIC methods are sufficient to employ a model comparison, they are more commonly used compared with BE.

• Comparison of different dark energy models.

In [46], based on the observational data of SNIa, BAO and CMB, Davis *et al.* scrutinized and compared a number of dark energy models by using AIC and BIC. They found that the ΛCDM model almost achieves the best fit of all the models despite its economy of parameters. A series of models, including the XCDM model with constant EOS of dark energy w, the Cardassian expansion model and the CPL parametrization, can also provide comparably good fits but have more free parameters. The DGP model and the standard CG model with $\alpha = 1$ are clearly disfavored. Based on their AIC and BIC values, one can determine the "rank" of these dark energy models (see Figure 18.11). From the figure, it is clear that the ΛCDM model is best favored, while the DGP model and the standard CG model are strongly disfavored.

By comparing the values of IC, similar results are obtained by Szydlowski *et al.* [209], A. Kurek *et al.* [210], Li *et al.* [63] and Wei *et al.* [65]. In the above works, the authors all found that the one-parameter flat ΛCDM performs best in the set of models considered in the context. These results further solidified the status of the ΛCDM scenario as the standard paradigm in modern cosmology, although there are still some puzzling conflicts between ΛCDM predictions and current observations [219].

To the contrary of the good performance of the ΛCDM model, another common results of these works are that the DGP model and the standard CG model usually perform badly and rank worst in the dark energy models considered thereof. So these two models are faced with high crisis when confronted with observations. Besides the DGP and the CG model, some other models like the ADE and RDE models also performed badly in a model comparison. For example, in [63], Li *et al.* obtained the following results of $\chi^2_{\min}$s from the combination of Constitution+BAO+CMB+H_0 data,

$$\chi^2_{\Lambda\text{CDM}} = 468.461, \quad \chi^2_{\text{RDE}} = 493.772,$$
$$\chi^2_{\text{ADE}} = 503.039, \quad \chi^2_{\text{DGP}} = 530.443.$$ (18.56)

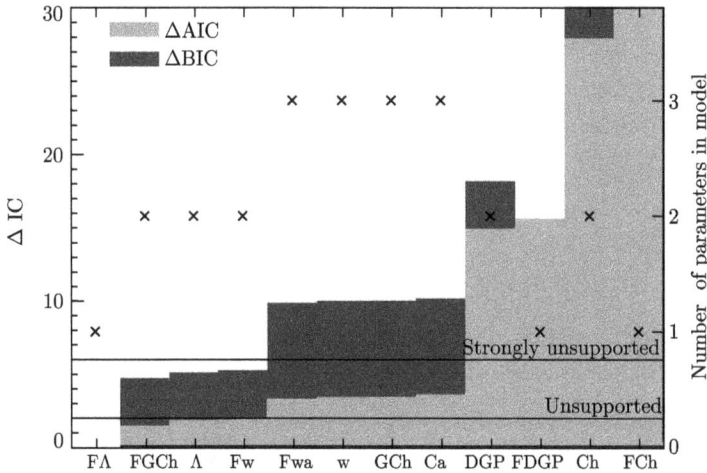

Figure 18.11   Graphical representation of the results of the IC values of some popular dark energy models. ΔAIC and ΔBIC are represented by the light and dark grey bars, respectively. From left to right, the models are given in order of increasing ΔAIC. The crosses mark the number of free parameters in each model (*right-hand ordinate*). The "unsupported" or "strongly unsupported" lines stand for a ΔBIC of 2 and 6. Clearly, the flat ΛCDM is the most preferred model. From [46]. Reproduced by permission of Tamara Davis and the AAS.

Compared with the ΛCDM model, the last three models have much larger BIC values,

$$\Delta\text{BIC}_{\text{RDE}} = 31.308\,, \quad \Delta\text{BIC}_{\text{ADE}} = 34.578\,,$$
$$\Delta\text{BIC}_{\text{DGP}} = 61.982\,. \tag{18.57}$$

What should be mentioned is that a somewhat different result was obtained by Sollerman *et al.* in [174], where they found that the flat DGP model performed even better than the flat ΛCDM model from first-year SDSS-II SNIa dataset [176] analyzing using the MLCS2k2 light-curve fitter. Notice that in this work, the authors also took two kinds of LTB models into consideration, which were not included in the model-comparison by the numerical studies listed above. In addition, they showed that the extra parameters required by these two models are not supported by the IC tests (their IC values are $\sim$ 5–25 larger compared with ΛCDM model).

Some other models, such as the general DGP model, the HDE model and the parameterizations like XCDM and CPL, can also provide rather good fits to the observational data. From current observational data it is hard to discriminate these models. Since they have more free parameters, these models are all less favored by the ΛCDM under the IC tests, indicating that given the current quality of the data there is no reason to prefer more complex models. The resulted ΔBIC values (with the ΛCDM model as a reference) of these models given in [63] are (in a flat universe)

$$\Delta BIC_{XCDM} = 5.862 \, , \quad \Delta BIC_{GDGP} = 5.897 \, , \qquad (18.58)$$

$$\Delta BIC_{CPL} = 11.195 \, , \quad \Delta BIC_{HDE} = 8.048 \, . \qquad (18.59)$$

Here the GDGP model refers to the generalized DGP model described by Eq. (18.30). Notice that the CPL model does not achieve a good performance due to its large number of free parameters. Another interesting phenomenon is that most of the above models (like the HDE model) can reduce to the ΛCDM model. From their best-fit parameters, it is found that they do tend to collapse to ΛCDM model [46, 63]. Therefore, the current observational data are still too limited to distinguish which theoretical model is better.

References

[1] I. Zlatev, L. M. Wang, and P. J. Steinhardt, Phys. Rev. Lett. **82** (1999) 896.
[2] R. R. Caldwell, R. Dave, and P. J. Steinhardt, Phys. Rev. Lett. **80** (1998) 1582.
[3] L. M. Wang and P. J. Steinhardt, ApJ. **508** (1998) 483.
[4] A. A. Starobinsky, JETP Lett. **68** (1998) 757.
[5] T. Nakamura and T. Chiba, MNRAS **306** (1999) 696.
[6] T. D. Saini *et al.*, Phys. Rev. Lett. **85** (2000) 1162.
[7] T. Chiba and T. Nakamura, Phys. Rev. D **62** (2000) 121301.
[8] C. P. Ma *et al.*, ApJ. **521** (1999) L1; Z. Haiman and J. J. Mohr, ApJ. **553** (2000) 545; P. Brax, J. Martin, and A. Riazuelo, Phys. Rev. D **62** (2000) 103505; B. Barger and D. Marfatia, Phys. Lett. B **498** (2001) 67; P. S. Corasaniti and E. J. Copeland, Phys. Rev. D **65** (2002) 043004; R. Bean and A. Melchiorri, Phys. Rev. D **65** (2002) 041302; B. F. Gerke and G. Efstathiou, MNRAS **335** (2002) 33; P. S. Corasaniti and E. J. Copeland, Phys. Rev. D **67** (2003) 063501; E. D. Pietro and J. F. Claeskens, MNRAS **341** (2003) 1299.
[9] V. Sahni, Class. Quant. Grav. **19** (2002) 3435.
[10] V. Sahni and A. Starobinsky, Int. J. Mod. Phys. D **15** (2006) 2105.
[11] S. Perlmutter *et al.*, ApJ. **517** (1999) 565.
[12] M. Hamuy *et al.*, AJ. **112** (1996) 2391.
[13] J. Simon, L. Verde, and R. Jimenez, Phys. Rev. D **71** (2005) 123001.

[14] Z. K. Guo, N. Ohta, and Y. Z. Zhang, Phys. Rev. D **72** (2005) 023504; Z. K. Guo, N. Ohta, and Y. Z. Zhang, Mod. Phys. Lett. A **22** (2007) 883; M. Sahlen, A. R. Liddle, and D. Parkinson, Phys. Rev. D **75** (2007) 023502.

[15] E. J. Copeland, M. Sami, and S. Tsujikawa, Int. J. Mod. Phys. D **15** (2006) 1753.

[16] R. Bean, S. M. Carroll, and M. Trodden, arXiv:astro-ph/0510059.

[17] T. Padmanabhan, Curr. Sci. **88** (2005) 1057.

[18] R. Lazkoz, S. Nesseris, and L. Perivolaropoulos, JCAP **0511** (2005) 010.

[19] G. Gupta, S. Panda and A. A. Sen, arXiv:1108.1322.

[20] V. B. Johri, Phys. Rev. D **70** (2004) 041303; S. Lee, G. C. Liu, and K. W. Ng, Phys. Rev. D **73** (2006) 083516; X. Zhang, Phys. Rev. D **74** (2006) 103505; R. Crittenden, E. Majerotto, and F. Piazza, Phys. Rev. Lett. **98** (2007) 251301; X. Zhang, Phys. Lett. B **648** (2007) 1; C. Schimd *et al.*, Astron. Astrophys. **463** (2007) 405; E. V. Linder, Gen. Rel. Grav. **40** (2008) 329; I. P. Neupane and C. Scherer, JCAP **0805** (2008) 009; B. J. Li, D. F. Mota, and J. D. Barrow, ApJ. **728** (2011) 109; H. Wei, Nucl. Phys. B **845** (2011) 381; O. Luongo and H. Quevedo, arXiv:1104.4758; A. Sheykhi, arXiv:1106.5697.

[21] C. Armendariz-Picon, JCAP **0407** (2004) 007.

[22] T. Padmanabhan, AIP Conf. Proc. **861** (2006) 179.

[23] V. V. Kiselev, Class. Quant. Grav. **21** (2004) 3323; H. Wei and R. G. Cai, Phys. Rev. D **73**, (2006) 083002; H. Wei and R. G. Cai, JCAP **0709** (2007) 015; J. B. Jimenez and A. L. Maroto, Phys. Rev. D **78** (2008) 063005; T. S. Koivisto and D. F. Mota, JCAP **0808** (2008) 021; K. Bamba, S. Nojiri, and S. D. Odintsov, Phys. Rev. D **77** (2008) 123532; J. B. Jimenez, R. Lazkoz, and A. L. Maroto, arXiv:0904.0433; V. A. De Lorenci, Class. Quant. Grav. **27** (2010) 065007.

[24] Y. Zhang, T. Y. Xia, and W. Zhao, Class. Quant. Grav. **24** (2007) 3309; T. Y. Xia and Y. Zhang, Phys. Lett. B **656** (2007) 19; S. Wang, Y. Zhang, and T. Y. Xia, JCAP **10** (2008) 037.

[25] E. Komatsu *et al.*, arXiv:1001.4538.

[26] R. Amanullah *et al.*, ApJ. **716** (2010) 712.

[27] W. J. Percival *et al.*, MNRAS **401** (2010) 2148.

[28] B. Feng *et al.*, Phys. Lett. B **634** (2006) 101; S. Nesseris and L. Perivolaropoulos, JCAP **0701** (2007) 018; H. Li *et al.*, Phys. Lett. B **658** (2008) 95.

[29] L. Amendola, Phys. Rev. D **62** (2000) 043511.

[30] C. Wetterich, Astron. Astrophys. **301** (1995) 321.

[31] L. Amendola *et al.*, ApJ. **583** (2003) L53; L. P. Chimento *et al.*, Phys. Rev. D **67** (2003) 083513; L. Amendola and C. Quercellini, Phys. Rev. D **68** (2003) 023514; L. Amendola, MNRAS **342** (2003) 221; G. Olivares, F. Atrio-Barandela, and D. Pavon, Phys. Rev. D **71** (2005) 063523; J. H. He and B. Wang, JCAP **0806** (2008) 010; J. Q. Xia, Phys. Rev. D **80** (2009) 103514; J. Valiviita, R. Maartens, and E. Majerotto, MNRAS **402** (2010) 2355; M. J. Mortonson, W. Hu, and D. Huterer, Phys. Rev. D **83** (2011) 023015.

[32] L. Amendola and D. Tocchini-Valentini, Phys. Rev. D **66** (2002) 043528.
[33] A. V. Maccio *et al.*, Phys. Rev. D **69** (2004) 123516.
[34] Z. K. Guo, N. Ohta, and S. Tsujikawa, Phys. Rev. D **76** (2007) 023508.
[35] R. Bean *et al.*, Phys. Rev. D **78** (2008) 123514.
[36] P. Astier *et al.*, Astron. Astrophys. **447** (2006) 31; S. Baumont *et al.*, Astron. Astrophys. **491** (2008) 567.
[37] C. L. Bennett *et al.*, ApJS. **148** (2003) 1; D. N. Spergel *et al.*, ApJS. **148** (2003) 175; D. N. Spergel *et al.*, ApJS. **170** (2007) 377; L. Page *et al.*, ApJS. **170** (2007) 335.
[38] D. J. Eisenstein *et al.*, ApJ. **633** (2005) 560.
[39] R. G. Cai and A. Z. Wang, JCAP **0503** (2005) 002; T. Koivisto, Phys. Rev. D **72** (2005) 043516; X. Zhang, Mod. Phys. Lett. A **20** (2005) 2575; X. Zhang, Phys. Lett. B **611** (2005) 1; M. Manera and D. F. Mota, MNRAS **371** (2006) 1373; G. Olivares, F. A. Barandela, and D. Pavon, Phys. Rev. D **74** (2006) 043521; J. H. He, B. Wang, and Y. P. Jing, JCAP **0907** (2009) 030; J. H. He, B. Wang, and P. J. Zhang, Phys. Rev. D **80** (2009) 063530; X. M. Chen *et al.*, Phys. Lett. B **695** (2011) 30; G. Caldera-Cabral, R. Maartens, and B. M. Schaefer, JCAP **0907** (2009) 027; E. Abdalla *et al.*, Phys. Lett. B **673** (2009) 107; A. Coc *et al.*, Phys. Rev. D **79** (2009) 103512; M. Baldi *et al.*, MNRAS **403** (2010) 1684; J. H. He, B. Wang, and E. Abdalla, arXiv:1012.3914.
[40] A. Y. Kamenshchik, U. Moschella, and V. Pasquier, Phys. Lett. B **511** (2001) 265.
[41] N. Bilic, G. B. Tupper, and R. D. Viollier, Phys. Lett. B **535** (2002) 17.
[42] M. C. Bento, O. Bertolami, and A. A. Sen, Phys. Rev. D **66** (2002) 043507.
[43] P. T. Silva and O. Bertolami, ApJ. **599** (2003) 829; M. Makler *et al.*, Phys. Lett. B. **555** (2003) 1; M. C. Bento, O. Bertolami, and A. A. Sen, Phys. Rev. D **67** (2003) 063003; M. C. Bento, O. Bertolami, and A. A. Sen, Phys. Lett. B **575** (2003) 172; A. Dev, J. S. Alcaniz, and D. Jain, Phys. Rev. D **67** (2003) 023515; L. Amendola *et al.*, JCAP **0307** (2003) 005; R. Colistete Jr *et al.*, Int. J. Mod. Phys. D **13** (2004) 669; O. Bertolami *et al.*, MNRAS **353** (2004) 329; T. Multamaki, M. Manera, and E. Gaztanaga, Phys. Rev. D **69** (2004) 023004; A. Dev, D. Jain, and J. S. Alcaniz, Astron. Astrophys. **417** (2004) 847; Y. G. Gong, JCAP **0503** (2005) 007; R. J. Colistete, J. C. Fabris, and S. V. B. Goncalves, Int. J. Mod. Phys. D **14** (2005) 775; J. S. Alcaniz and J. A. S. Lima, ApJ. **618** (2005) 16; M. Biesiada, W. Godlowski, and M. Szydlowski, ApJ. **622** (2005) 28; X. Zhang, F. Q. Wu, and J. F. Zhang, JCAP **0601** (2006) 003; O. Bertolami and P. T. Silva, MNRAS **365** (2006) 1149; T. Giannantonio and A. Melchiorri, Class. Quant. Grav. **23** (2006) 4125; P. X. Wu and H. W. Yu, JCAP **0703** (2007) 015; Z. Li, P. Wu, and H. Yu, JCAP **0909** (2009) 017.
[44] H. Sandvik *et al.*, Phys. Rev. D **69** (2004) 123524.
[45] Z. H. Zhu, Astron. Astrophys. **423** (2004) 421.
[46] T. M. Davis *et al.*, ApJ. **666** (2007) 716.
[47] R. A. Daly and S. G. Djorgovski, ApJ. **597** (2003) 9.

[48] X. H. Meng, J. Ren, and M. G. Hu, Commun. Theor. Phys. **47** (2007) 379.
[49] J. Ren and X. H. Meng, Phys. Lett. B **633** (2006) 1.
[50] M. Makler, S. Q. de Oliveira, and I. Waga, Phys. Rev. D **68** (2003) 123521.
[51] M. Colless *et al.*, arXiv:astro-ph/0306581.
[52] L. M. G. Beca *et al.*, Phys. Rev. D **67** (2003) 101301; M. C. Bento, O. Berto-lami, and A. A. Sen, Phys. Rev. D **70** (2004) 083519; V. Gorini *et al.*, JCAP **0802** (2008) 016; J. C. Fabris *et al.*, Phys. Lett. B, **694** (2011) 289.
[53] C. G. Park *et al.*, Phys. Rev. D **81** (2010) 063532.
[54] M. Li, Phys. Lett. B **603** (2004) 1.
[55] Q. G. Huang and Y. G. Gong, JCAP **0408** (2004) 006.
[56] X. Zhang and F. Q. Wu, Phys. Rev. D **72** (2005) 043524.
[57] Z. L. Yi and T. J. Zhang, Mod. Phys. Lett. A **22** (2007) 41.
[58] Q. G. Huang and M. Li, JCAP **0408** (2004) 013; J. Y. Shen *et al.*, Phys. Lett. B **609** (2005) 200; Z. Chang, F. Q. Wu, and X. Zhang, Phys. Lett. B **633** (2006) 14; X. Zhang and F. Q. Wu, Phys. Rev. D **76** (2007) 023502; M. R. Setare, J. F. Zhang, and X. Zhang, JCAP **0703** (2007) 007; Y. T. Wang and L. X. Xu, Phys. Rev. D **81** (2010) 082523; X. Zhang, Phys. Lett. B **683** (2010) 81; S. del Campo, J. C. Fabris, and R. Herrera, arXiv:1103.3441.
[59] M. Li *et al.*, JCAP **0906** (2009) 036.
[60] Y. Z. Ma, Y. Gong, and X. L. Chen, Eur. Phys. J. C **60** (2009) 303.
[61] Y. Z. Ma, AIP Conf. Proc. **1166** (2009) 44.
[62] Y. G. Gong, B. Wang, and Y. Z. Zhang, Phys. Rev. D **72** (2005) 043510.
[63] M. Li, X. D. Li, and X. Zhang, Sci. China Phys. Mech. Astron. **53** (2010) 1631.
[64] M. Hicken *et al.*, ApJ. **700** (2009) 1097; M. Hicken *et al.*, ApJ. **700** (2009) 331.
[65] H. Wei, JCAP **1008** (2010) 020.
[66] B. Wang, Y. G. Gong, and E. Abdalla, Phys. Lett. B **624** (2005) 141.
[67] B. Wang, C. Y. Lin, and E. Abdalla, Phys. Lett. B **637** (2006) 357.
[68] A. Barreira and P. P. Avelino, Phys. Rev. D **83** (2011) 103001
[69] B. Wang *et al.*, Phys. Lett. B **662** (2008) 1.
[70] J. L. Cui and X. Zhang, Phys. Lett. B **690** (2010) 233.
[71] D. Pavon and W. Zimdahl, Phys. Lett. B **628** (2005) 206; H. M. Sadjadi and M. Honardoost, Phys. Lett. B **647** (2007) 231; M. R. Setare, Phys. Lett. B **654** (2007) 1; W. Zimdahl and D. Pavon, Class. Quant. Grav. **24** (2007) 5461; J. Zhang, X. Zhang, and H. Liu, Phys. Lett. B **659** (2008) 26.
[72] Y. Z. Ma, Y. Gong, and X. L. Chen, Eur. Phys. J. C **69** (2010) 509.
[73] C. Feng *et al.*, JCAP **0709** (2007) 005.
[74] M. Li *et al.*, JCAP **0912** (2009) 014.
[75] R. G. Cai, Phys. Lett. B **657** (2007) 228.
[76] H. Wei and R. G. Cai, Phys. Lett. B **660** (2008) 113.
[77] H. Wei and R. G. Cai, Eur. Phys. J. C **59** (2009) 99; I. P. Neupane, Phys. Lett. B **673** (2009) 111; I. P. Neupane, Phys. Rev. D **76** (2006) 123006.
[78] C. J. Gao *et al.*, Phys. Rev. D **79** (2009) 043511.

[79] L. X. Xu, W. B. Li, and J. B. Lu, Mod. Phys. Lett. A **24** (2009) 1355.

[80] X. Zhang, Phys. Rev. D **79** (2009) 103509.

[81] C. J. Feng, Phys. Lett. B **670** (2008) 231.

[82] H. Wei and R. G. Cai, Phys. Lett. B **663** (2008) 1.

[83] K. Y. Kim, H. W. Lee, and Y. S. Myung, Phys. Lett. B **660** (2008) 118.

[84] G. R. Dvali, G. Gabadadze, and M. Porrati, Phys. Lett. B **485** (2000) 208.

[85] G. Dvali and M. S. Turner, astro-ph/0301510.

[86] D. Rubin *et al.*, ApJ. **695** (2009) 391.

[87] W. J. Fang *et al.*, Phys. Rev. D **78** (2008) 103509.

[88] Z. K. Guo *et al.*, ApJ. **646** (2006) 1.

[89] M. Fairbairn and A. Goodbar, Phys. Lett. B **642** (2006) 432.

[90] R. Maartens and E. Majerotto, Phys. Rev. D **74** (2006) 023004.

[91] U. Alam and V. Sahni, Phys. Rev. D **73** (2006) 084024.

[92] J. Q. Xia, Phys. Rev. D **79** (2009) 103527.

[93] A. Lue, R. Scoccimarro, and G. Starkman, Phys. Rev. D **69** (2004) 124015.

[94] A. Lue, Phys. Rept. **423** (2006) 1.

[95] I. Sawicki and S. M. Carroll, arXiv:astro-ph/0510364; K. Yamamoto *et al.*, Phys. Rev. D **74** (2006) 063525; J. P. Uzan, Gen. Rel. Grav. **39** (2007) 307; H. Wei, Phys. Lett. B **664** (2008) 1; L. Lombriser *et al.*, Phys. Rev. D **80** (2009) 063536; K. C. Chan and R. Scoccimarro, Phys. Rev. D **80** (2009) 104005.

[96] Y. S. Song, I. Sawicki, and W. Hu, Phys. Rev. D **75** (2007) 064003.

[97] A. Cardoso *et al.*, Phys. Rev. D **77** (2008) 083512.

[98] W. J. Fang, W. Hu, and A. Lewis, Phys. Rev. D **78** (2008) 087303.

[99] F. Schmidt, Phys. Rev. D **80** (2009) 043001.

[100] S. Tsujikawa, arXiv:1004.1493.

[101] T. P. Sotiriou and V. Faraoni, Rev. Mod. Phys. **82** (2010) 451.

[102] S. Nojiri and S. D. Odintsov, arXiv:hep-th/0601213.

[103] A. D. Felice and S. Tsujikawa, Living Rev. Rel. **13** (2010) 3.

[104] S. Capozziello, M. D. Laurentis, and V. Faraoni, arXiv:0909.4672.

[105] S. Nojiri and S. D. Odintsov, Phys. Rept. **505** (2011)59.

[106] S. Capozziello, Int. J. Mod. Phys. D **11** (2002) 483.

[107] S. Capozziello *et al.*, Int. J. Mod. Phys. D **12** (2003) 1969.

[108] S. Nojiri and S. D. Odintsov, Phys. Rev. D **68** (2003) 123512.

[109] A. D. Dolgov and M. Kawasaki, Phys. Lett. B **573** (2003) 1; V. Faraoni, Phys. Rev. D **74** (2006) 104017.

[110] G. J. Olmo, Phys. Rev. Lett. **95** (2005) 261102; G. J. Olmo, Phys. Rev. D **72** (2005) 083505; I. Navarro and K. V Acoleyen, JCAP **0702** (2007) 022; A. L. Erickcek, T. L. Smith, and M. Kamionkowski, Phys. Rev. D **74** (2006) 121501; T. Chiba, T. L. Smith, and A. L. Erickcek, Phys. Rev. D **75** (2007) 124014.

[111] G. Cognola *et al.*, JCAP **0502** (2005) 010.

[112] V. Faraoni, Phys. Rev. D **70** (2004) 04437.

[113] W. Hu and I. Sawicki, Phys. Rev. D **76** (2007) 064004.

[114] V. Muller, H. J. Schmidt, and A. A. Starobinsky, Phys. Lett. B **202** (1988) 198.

[115] L. Amendola et al., Phys. Rev. D **75** (2007) 083504.

[116] A. A. Starobinsky, J. Exp. Theor. Phys. Lett. **86** (2007) 157.

[117] L. Amendola, D. Polarski, and S. Tsujikawa, Phys. Rev. Lett. **98** (2007) 131302; L. Amendola, D. Polarski and S. Tsujikawa, Int. J. Mod. Phys. D **16** (2007) 1555.

[118] S. Nojiri and S. D. Odintsov, Phys. Rev. D **74** (2006) 086005.

[119] S. Nojiri and S. D. Odintsov, Phys. Lett. B **657** (2007) 238; S. Nojiri and S. D. Odintsov, Phys. Rev. D **77** (2008) 026007; G. Cognola et al., Phys. Rev. D **77** (2008) 046009.

[120] Y. S. Song, W. Hu, and I. Sawicki, Phys. Rev. D **75** (2007) 044004.

[121] Y. S. Song, H. Peiris, and W. Hu, Phys. Rev. D **76** (2007) 063517.

[122] P. J. Zhang et al., Phys. Rev. Lett. **99** (2007) 141302; R. Bean et al., Phys. Rev. D **75** (2007) 064020; S. Nojiri and S. D. Odintsov, J. Phys. Conf. Ser. **66** (2007) 012005; T. Faulkner et al., Phys. Rev. D **76** (2007) 063505; L. M. Sokolowski, Class. Quant. Grav. **24** (2007) 3713; P. J. Zhang, Phys. Rev. D **76** (2007) 024007; S. Tsujikawa et al., Phys. Rev. D **80** (2009) 084044; A. De Felice, S. Mukohyama, and S. Tsujikawa, Phys. Rev. D **82** (2010) 023524.

[123] B. Jain and J. Khoury, Annals Phys. **325** (2010) 1479.

[124] F. Schmidt et al., Phys. Rev. D **79** (2009) 083518; F. Schmidt, A. Vikhlinin, and W. Hu, Phys. Rev. D **80** (2009) 082505; F. Schmidt, Phys. Rev. D **81** (2010) 103002.

[125] L. Lombriser et al., arXiv:1003.3009.

[126] S. Tsujikawa and T. Tatekawa, Phys. Lett. B **665** (2008) 325; F. Schmidt, Phys. Rev. D **78** (2008) 043002; A. Borisov and B. Jain, Phys. Rev. D **79** (2009) 103506; T. Narikawa and K. Yamamoto, Phys. Rev. D **81** (2010) 043528.

[127] S. Camera, A. Diaferio and V. F. Cardone, JCAP **07** (2011) 016.

[128] J. S. B. Wyithe and A. Loeb, ApJ. **588** (2003) L59; R. Cen, ApJ. **591** (2003) L5; Z. Haiman and G. P. Holder, ApJ. **595** (2003) 1; M. Zaldarriaga, S. R. Furlanetto, and L. Hernquist, ApJ. **608** (2004) 622; S. Furlanetto, S. P. Oh, and F. Briggs, Phys. Rept. **433** (2006) 181; M. McQuinn et al., MNRAS **377** (2007) 1043; M. McQuinn et al., ApJ. **653** (2006) 815; J. R. Pritchard and A. Loeb, Phys. Rev. D **78** (2008) 103511; X. C. Mao and X. Wu, ApJ. **673** (2008) L107; X. L. Chen and J. Miralda-Escude, ApJ. **684** (2008) 18.

[129] K. W. Masui et al., Phys. Rev. D **81** (2010) 062001.

[130] J. K. Webb et al., Phys. Rev. Lett. **82** (1999) 884; J. K. Webb et al., Phys. Rev. Lett. **87** (2001) 091301; J. K. Webb et al., arXiv:1008.3907.

[131] K. A. Olive, M. Peloso, and J. P. Uzan, Phys. Rev. D **83** (2011) 043509; T. Chiba and M. Yamaguchi, JCAP **1103** (2011) 044; K. Bamba, S. Nojiri and S. D. Odintsov, arXiv:1107.2538.

[132] J. P. Uzan, Gen. Rel. Grav. **39** (2007) 307.

[133] J. P. Uzan and F. Bernardeau, Phys. Rev. D **64** (2001) 083004; H. Oyaizu, M. Lima, and W. Hu, Phys. Rev. D **78** (2008) 123524; K. Koyama, A. Taruya, and T. Hiramatsu, Phys. Rev. D **79** (2009) 123512; P. J. Zhang,

Phys. Rev. D **73** (2006) 123504; K. N. Ananda, S. Carloni, and P. K. S. Dunsby, Class. Quant. Grav. **26** (2009) 235018; T. Giannantonio *et al.*, JCAP **1004** (2010) 030; S. Capozziello and M. Francaviglia, Gen. Rel. Grav. **40** (2008) 357; B. Jain and P. J. Zhang, Phys. Rev. D **78** (2008) 063503; M. C. Martino and R. K. Sheth, arXiv:0911.1829; I. Tereno, E. Semboloni, and T. Schrabback, arXiv:1012.5854; J. P. Uzan, Gen. Rel. Grav. **42** (2010) 2219; S. Unnikrishnan, S. Thakur, and T. R. Seshadri, arXiv:1106.6353.

[134] M. C. B. Abdalla, S. Nojiri, and S. D. Odintsov, Class. Quant. Grav. **22** (2005) L35.

[135] F. Briscese *et al.*, Phys. Lett. B **646** (2007) 105.

[136] A. V. Frolov, Phys. Lett. **101** (2008) 061103; T. Kobayashi and K. I. Madeda, Phys. Rev. D **78** (2008) 064019.

[137] S. Nojiri and S. D. Odintsov, Phys. Rev. D **78** (2008) 046006; K. Bamba, S. Nojiri, and S. D. Odintsov, JCAP **0810** (2008) 045.

[138] A. Dev *et al.*, Phys. Rev. D **78** (2008) 083515; T. Kobayashi and K. I. Maeda, Phys. Rev. D **79** (2009) 024009; S. Appleby, R. Battye, and A. Starobinsky, JCAP **1006** (2010) 005; E. Babichev and D. Langlois, Phys. Rev. D **81** (2010) 124051.

[139] M. Amarzguioui *et al.*, Astron. Astrophys. **454** (2006) 707.

[140] T. Koivisto, Phys. Rev. D **73** (2006) 083517.

[141] M. Tegmark *et al.*, ApJ. **606** (2004) 702.

[142] E. E. Flanagan, Phys. Rev. Lett. **92** (2004) 071101; E. E. Flanagan, Class. Quant. Grav. **21** (2004) 417; E. E. Flanagan, Class. Quant. Grav. **21** (2004) 3817.

[143] X. H. Meng and P. Wang, Class. Quant. Grav. **20** (2003) 4949; X. H. Meng and P. Wang, arXiv:astro-ph/0308284; X. H. Meng and P. Wang, Phys. Lett. B **584** (2004) 1; X. H. Meng and P. Wang, Gen. Rel. Grav. **36** (2004) 1947; X. H. Meng and P. Wang, Class. Quant. Grav. **21** (2004) 951.

[144] E. Barausse, T. P. Sotiriou, and J. C. Miller, Class. Quant. Grav. **25** (2008) 105008; E. Barausse, T. P. Sotiriou, and J. C. Miller, Class. Quant. Grav. **25** (2008) 062001; G. J. Olmo, Phys. Rev. D **78** (2008) 104026; G. J. Olmo and P. Singh, JCAP **0901** (2009) 030; C. Barragan, G. J. Olmo, and H. Sanchis-Alepuz, Phys. Rev. D **80** (2009) 024016; B. Li, D. F. Mota, and D. J. Shaw, Class. Quant. Grav. **26** (2009) 055018.

[145] K. S. Stelle, Gen. Rel. Grav. **9** (1978) 353; N. H. Barth and S. M. Christensen, Phys. Rev. D **28** (1983) 1876; A. De Felice, M. Hindmarsh, and M. Trodden, JCAP **0608** (2006) 005; G. Calcagni, B. de Carlos, and A. De Felice, Nucl. Phys. B **752** (2006) 404.

[146] T. Koivisto and D. F. Mota, Phys. Lett. B **644** (2007) 104.

[147] T. Koivisto and D. F. Mota, JCAP **0701** (2007) 006.

[148] B. Li, J. D. Barrow, and D. F. Mota, Phys. Rev. D **75** (2007) 023520.

[149] A. De Felice, D. F. Mota, and S. Tsujikawa, Phys. Rev. D **81** (2010) 023532.

[150] C. Brans and R. H. Dicke, Phys. Rev. **124** (1961) 925.

[151] C. M. Will, Living Rev. Rel. **9** (2005) 3.
      URL: http://www.livingreviews.org/lrr-2006-3.

[152] L. Amendola, Phys. Rev. D **60** (1999) 043501.
[153] F. Q. Wu *et al.*, Phys. Rev. D **82** (2010) 083002; F. Q. Wu and X. L. Chen, Phys. Rev. D **82** (2010) 083003.
[154] A. Riazuelo and J. P. Uzan, Phys. Rev. D **62** (2000) 083506; A. Riazuelo and J. P. Uzan, Phys. Rev. D **66** (2002) 023525; A. Coc *et al.*, Phys. Rev. D **73** (2006) 083525; C. Schimd, J. P. Uzan, and A. Riazuelo, Phys. Rev. D **71** (2005) 083512.
[155] G. R. Bengochea and R. Ferraro, Phys. Rev. D **79** (2009) 124019.
[156] E. V. Linder, Phys. Rev. D **81** (2010) 127301.
[157] P. X. Wu and H. W. Yu, Phys. Lett. B **693** (2010) 415.
[158] G. R. Bengochea, Phys. Lett. B **695** (2011) 405.
[159] H. Wei, X. P. Ma, and H. Y. Qi, Phys. Lett. B **703** (2011) 74.
[160] H. Wei, H. Y. Qi, and X. P. Ma, arXiv:1108.0859.
[161] K. Bamba *et al.*, JCAP **1101** (2011) 021.
[162] S. H. Chen *et al.*, Phys. Rev. D **83** (2011) 023508.
[163] J. B. Dent, S. Dutta, and E. N. Saridakis, JCAP **1101** (2011) 009.
[164] R. Zheng and Q. G. Huang, JCAP **1103** (2011) 002.
[165] B. Li, T. P. Sotiriou, and J.D. Barrow, arXiv:1010.1041.
[166] C. H. Chuang, J. A. Gu, and W. Y. Hwang, Class. Quant. Grav. **25** (2008) 175001.
[167] A. Paranjape and T. P. Singh, Class. Quant. Grav. **23** (2006) 6955.
[168] G. Lemaître, Annales Soc. Sci. Brux. Ser. ISci. Math. Astron. Phys. A **53** (1933) 51.
[169] R. C. Tolman, PNAS **20** (1934) 169.
[170] H. Bondi, MNRAS **107** (1947) 410.
[171] H. Alnes, M. Amarzguioui, and O. Gron, Phys. Rev. D **73** (2006) 083519.
[172] K. Tomita, ApJ. **529** (2000) 38; K. Tomita, MNRAS **326** (2001) 287; M. N. Celerier, Astron. Astrophys. **353** (2000) 63.
[173] N. Mustapha, C. Hellaby, and G. F. R. Ellis, MNRAS **292** (1997) 817; M. N. Célériér, K. Bolejko, and A. Krasiéski, arXiv:0906.0905; E. W. Kolb and C. R. Lamb, arXiv:0911.3852; A. E. Romano, JCAP **1001** (2010) 004; A. E. Romano, JCAP **05** (2010) 020.
[174] J. Sollerman *et al.*, ApJ. **703** (2009) 1374.
[175] J. G. Bellido and T. Haugbolle, JCAP **4** (2008) 3.
[176] J. A. Holtzman *et al.*, AJ. **136** (2008) 2306; R. Kessler *et al.*, ApJS. **185** (2009) 32.
[177] H. Alnes and M. Amarzguioui, Phys. Rev. D **74** (2006) 103520.
[178] C. Quercellini, M. Quartin, and L. Amendola, Phys. Rev. Lett. **102** (2009) 151302.
[179] T. Biswas, A. Notari, and W. Valkenburg, JCAP **1011** (2010) 030.
[180] A. Moss, J. P. Zibin, and D. Scott, arXiv:1007.3725.
[181] M. Blomqvist and E. Mortsell, JCAP **1005** (2010) 6.
[182] J. G. Bellido and T. Haugboelle, JCAP **0809** (2008) 016.
[183] D. L. Wiltshire, Phys. Rev. Lett. **99** (2007) 251101.
[184] J. P. Zibin, A. Moss, and D. Scott, Phys. Rev. Lett. **101** (2008) 251303; J. P. Zibin, Phys. Rev. D **78** (2008) 043504.

[185] S. Nadathur and S. Sarkar, Phys. Rev. D **83** (2011) 063506.
[186] A. G. Riess *et al.*, ApJ. **730** (2011) 119.
[187] J. P. Uzan, C. Clarkson, and G. F. R. Ellis, Phys. Rev. Lett. **100** (2008) 191303; P. Dunsby *et al.*, JCAP **1006** (2010) 017.
[188] C. M. Yoo, K. I. Nakao, and M. Sasaki, JCAP **1007** (2010) 012.
[189] M. Quartin and L. Amendola, Phys. Rev. D **81** (2010) 043522.
[190] J. Goodman, Phys. Rev. D **52** (1995) 1821.
[191] R. R. Caldwell and A. Stebbins, Phys. Rev. Lett. **100** (2008) 191302.
[192] T. Clifton, P. G. Ferreira, and K. Land, Phys. Rev. Lett. **101** (2008) 131302.
[193] C. Clarkson, B. Bassett, and T. H. C. Lu, Phys. Rev. Lett. **101** (2008) 011301.
[194] T. Clifton, P. G. Ferreira, and J. Zuntz, JCAP **0907** (2009) 029.
[195] J. J. Jia and H. B. Zhang, JCAP **0810** (2008) 002.
[196] M. X. Lan *et al.*, Phys. Rev. D **82** (2010) 023516.
[197] T. Buchert, Gen. Rel. Grav. **40** (2008) 467.
[198] J. G. Bellido and T. Haugboelle, JCAP **0804** (2008) 003; S. February *et al.*, MNRAS **405** (2010) 2231; R. A. Vanderveld, E. E. Flanagan, and I. Wasserman, Phys. Rev. D **78** (2008) 083511; S. Alexander *et al.*, JCAP **0909** (2009) 025; K. Bolejko and J. S. B. Wyithe, JCAP **0902** (2009) 020; C. Clarkson and M. Regis, arXiv:1007.3443.
[199] S. Rasanen, arXiv:1102.0408.
[200] S. Rasanen, JCAP **0902** (2009) 011.
[201] S. Rasanen, JCAP **0402** (2004) 003.
[202] E. W. Kolb *et al.*, Phys. Rev. D **71** (2005) 023524.
[203] E. W. Kolb, S. Matarrese, and A. Riotto, New J. Phys. **8** (2006) 322.
[204] N. Li and D. J. Schwarz, Phys. Rev. D **78** (2008) 083531.
[205] M. Seikel and D. J. Schwarz, arXiv:0912.2308.
[206] J. Larena *et al.*, Phys. Rev. D **79** (2009) 083011.
[207] T. Buchert and M. Carfora, Class. Quant. Grav. **25** (2008) 195001; C. Clarkson, K. Ananda, and J. Larena, Phys. Rev. D **80** (2009) 083525; E. W. Kolb, V. Marra, and S. Matarrese, Gen. Rel. Grav. **42** (2010) 1399; O. Umeh, J. Larena, and C. Clarkson, arXiv:1011.3959; R. A. Sussman, arXiv:1102.2663.
[208] S. Nesseris and L. Perivolaropoulos, Phys. Rev. D **70** (2004) 043531.
[209] M. Szydlowski, A. Kurek, and A. Krawiec, Phys. Lett. B **642** (2006) 171.
[210] A. Kurek and M. Szydlowski, ApJ. **675** (2008) 1.
[211] Y. G. Gong and C. K. Duan, MNRAS **352** (2004) 847; M. Li, X. D. Li, and S. Wang, arXiv:0910.0717; S. Basilakos, M. Plionis, and J. A. S. Lima, Phys. Rev. D **82** (2010) 083517.
[212] G. Schwarz, The Annals of Statistics **6** (1978) 461.
[213] H. Akaike, IEEE Transactions on Automatic Control **19** (1974) 716.
[214] W. Godlowski and M. Szydlowski, Phys. Rev. Lett. B **623** (2005) 10; M. Biesiada, JCAP **0702** (2007) 003; J. Magueijo and R. D. Sorkin, MNRAS **377** (2007) L39.
[215] A. R. Liddle, MNRAS **351** (2004) L49.

[216] P. Mukherjee, D. R. Parkinson, and A. R. Liddle, ApJ. **638** (2006) L51.

[217] T. D. Saini, J. Weller, and S. L. Bridle, MNRAS **348** (2004) 603; A. R. Liddle *et al.*, Phys. Rev. D **74** (2006) 123506; O. Elgaroy and T. Multamaki, JCAP **0609** (2006) 002; P. Marshall, N. Rajgru, and A. Slosar, Phys. Rev. D **73** (2006) 067302.

[218] Y. Gong and X. L. Chen, Phys. Rev. D **76** (2007) 123007.

[219] L. Perivolaropoulos and A. Shafieloo, Phys. Rev. D **79** (2009) 123502; L. Perivolaropoulos, arXiv:0811.4684, arXiv:1002.3030, arXiv:1104.0539.

19
Dark Energy Reconstructions from Observational Data

In the last chapter, we have introduced some representative numerical works on the specific dark energy models. Due to the lack of a compelling fundamental theory to explain the dark energy, another route, the model-independent dark energy reconstructions, have drawn more and more attentions [1–9].

The dark energy reconstruction is a classic statistical inverse problem for the Hubble parameter

$$H(z) = H_0 \sqrt{\Omega_{m0}(1 + z)^3 + (1 - \Omega_{m0})f(z)}, \qquad (19.1)$$

where $f(z) \equiv \rho_{de}(z)/\rho_{de}(0)$ is the dark energy density function. Different reconstruction method will give different $f(z)$. An ideal dark energy reconstruction should be sufficiently versatile to accommodate a large class of dark energy models. The main target of a dark energy reconstruction is to detect the dynamical property of dark energy, i.e. determine whether the accelerating expansion is consistent with a cosmological constant.

To begin with one should choose an appropriate quantity characterizing dark energy. It is widely believed that the EOS of dark energy $w_{de} \equiv p_{de}/\rho_{de}$ holds essential clues for the nature of DE [10]. It should be mentioned that $w_{de}(z)$ is related with $f(z)$ through an integration [11–13]

$$f(z) = \exp\left(3 \int_0^z dz' \frac{1 + w_{de}(z')}{1 + z'}\right). \qquad (19.2)$$

Since the ΛCDM model always satisfies $w_{de} = -1$, the deviation from this constant EOS will reveal the variation of dark energy density. Therefore, most researchers have chosen to study dark energy by constraining w_{de} from observations.

However, in a series of works [11–16], Wang and collaborators argued that, due to the smearing effect [5] arising from the multiple integrals relating $w_{de}(z)$ to the luminosity distance of SN $d_L(z)$, it is difficult to constrain w_{de} using the SN data [4]. On the contrary, since using the dark energy density ρ_{de} can minimize the smearing effect by removing one integral, ρ_{de} can be constrained more tightly than w_{de} given the same observational data (see Figure 19.1 for details). It should be mentioned that there is still a debate on which quantity, w_{de} or ρ_{de}, is better in describing dark energy [17].

In addition to w_{de} and ρ_{de}, some other quantities, such as the deceleration parameter q [3, 18–20], the redshift of acceleration-deceleration transition [21], the statefinder diagnostic (r, s) [22], the jerk parameter j [23, 24] and the diagnostic Om [25, 26], can also provide very useful information for the study of dark energy. Although these quantities can accurately reconstruct some dark energy models, they have difficulty to discriminate between different models of dark energy [27]. So in this chapter, we will introduce the model-independent methods based on the reconstructions of w_{de} and ρ_{de}.

The model-independent dark energy reconstructions can be divided into four classes: (i) Specific Ansatz: assuming a specific parameterized form for $w_{de}(z)$ and estimating the associated parameters. (ii) Binned Parametrization: dividing the redshift range into different bins and using a simple local basis representation for $w_{de}(z)$ or $\rho_{de}(z)$. (iii) Polynomial Fitting: treating the dark energy density function $f(z) \equiv \rho_{de}(z)/\rho_{de}(0)$ as a free function of redshift and representing it by using the polynomial. (iv) Gaussian Process modeling: using a distribution over functions that can represent $w_{de}(z)$ and estimating the statistical properties thereof. These four classes of reconstruction methods and the related research works will be introduced in the following.

19.1   Specific Ansatz

The "specific ansatz" is the most popular approach currently. The key idea is assuming a specific parameterized form for $w_{de}(z)$ and estimating the associated parameters. A simple and widely used ansatz is the XCDM ansatz, in which the EOS of dark energy is a constant, i.e. $w_{de} = \text{const}$.

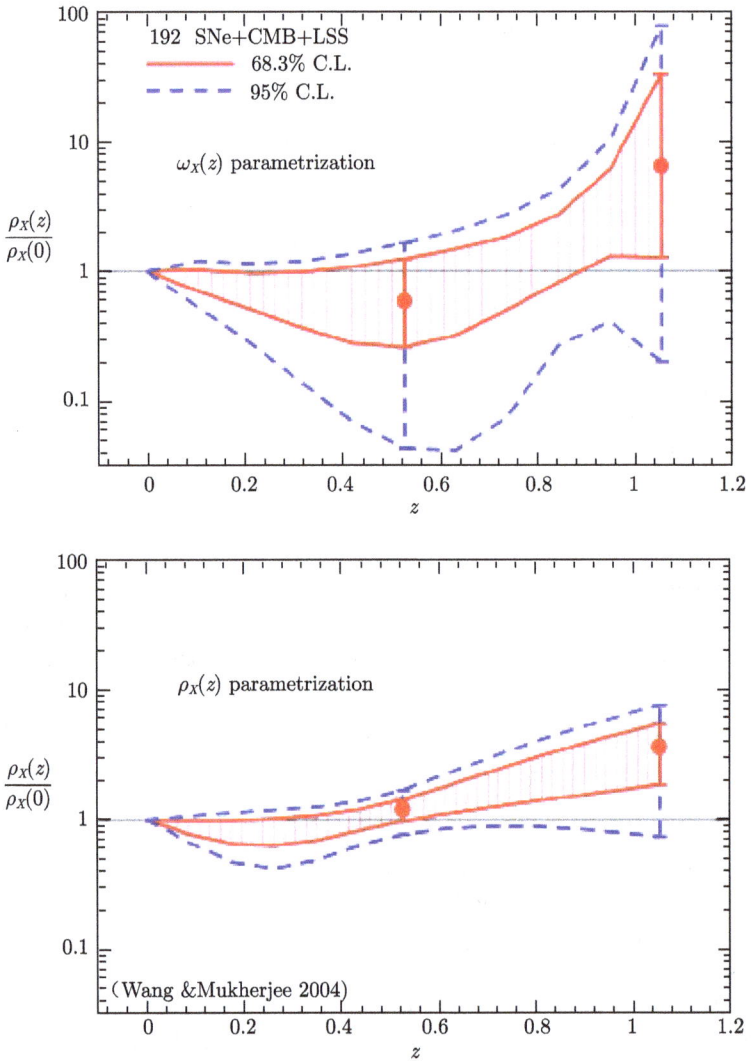

Figure 19.1   A comparison between the w_{de} parametrization and the ρ_{de} parametrization using the same observational data. The regions inside the solid and dashed lines correspond to 1σ and 2σ confidence regions, respectively. Clearly the ρ_{de} reconstruction can give much tighter constraint on dark energy compared with the w_{de} parametrization. Reprinted from [13], with permission from Yun Wang and Elsevier.

This yields a simple form of $f(z)$,

$$f(z) = (1+z)^{3(1+w_{de})}. \tag{19.3}$$

In [28], by combining the WMAP5 observations with BAO and SN data, Komatsu et al. obtained $w_{de} = -0.992^{+0.061}_{-0.062}$ at the 1σ CL, while in [29], a combination analysis of WMAP7+BAO+SN gave $w_{de} = -0.980^{+0.053}_{-0.053}$. A more recent constraint on w_{de} by the SCP team [30] also presented the consistent result (see Figure 19.2). So the current observations still favor $w_{de} = -1$ (i.e. ΛCDM model).

Besides, one can also assume that the EOS of dark energy is not a constant. The most popular parametrization with dynamical w_{de}, which assumes $w_{de}(z) = w_0 + w_a z/(1+z)$, was firstly proposed by Chevallier and Polarski [31], then was used to explore the expansion history of the universe by Linder [32]. So this ansatz is often called CPL parametrization. The corresponding $f(z)$ is given by [31, 33]

$$f(z) = (1+z)^{3(1+w_0+w_a)} \exp\left(-\frac{3w_a z}{1+z}\right). \tag{19.4}$$

Here w_0 denotes the value of the present EOS, while w_a denotes the variation of the EOS. Because of its bounded behavior at high redshift and high accuracy in reconstructing many scalar field EOS [32], the CPL parametrization has become one of the most popular methods to study dark energy [7, 34]. In Figure 19.3 we show the constraint on (w_0, w_a) in the CPL ansatz given by the Union2 SNIa dataset [30] and by the WMAP7 observations [29]. Obviously, the current data favor the result of $w_0 = -1$ and $w_a = 0$, which is consistent with the cosmological constant.

Although the CPL parametrization has many advantages, due to the fact that $|w(z)|$ grows rapidly and finally encounters divergence as z approaches -1, it cannot describe the future expansion history and thus loses the prediction ability for the fate of the Universe. In [36], Ma and Zhang proposed a divergence-free parameterization:

$$w(z) = w_0 + w_1 \left(\frac{\ln(2+z)}{1+z} - \ln 2\right). \tag{19.5}$$

For simplicity, we will call this model MZ parametrization. Thanks to the logarithm form, the MZ parametrization has well behaved, bounded behavior for both high redshifts and negative redshifts. In addition, the dark energy density $f(z)$ can also be obtained analytically. In Figure 19.4

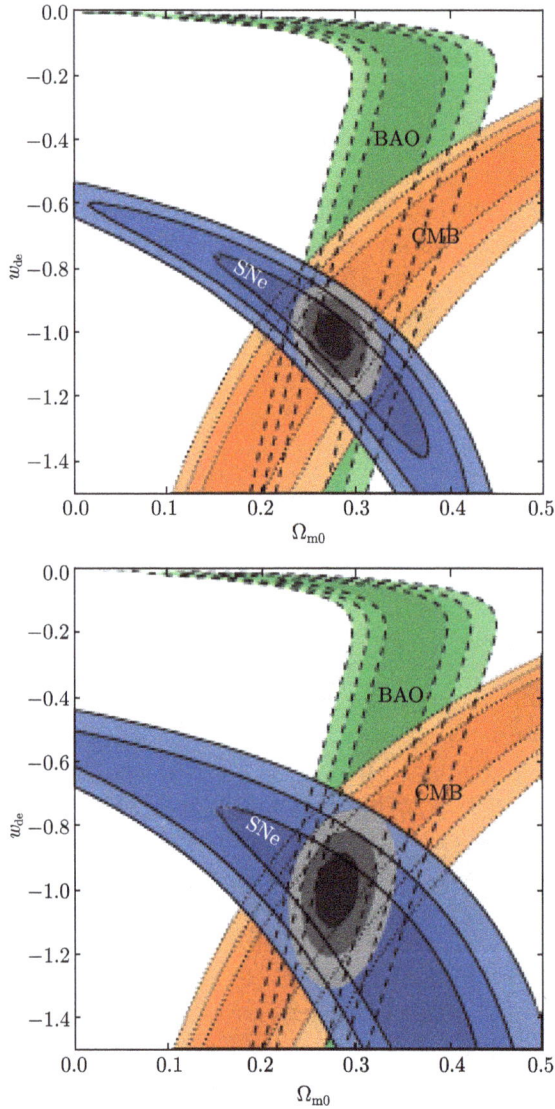

Figure 19.2   68.3%, 95.4%, and 99.7% confidence regions of the (Ω_{m0}, w_{de}) plane from SN combined with the constraints from BAO and CMB both without (upper panel) and with (lower panel) systematic errors. From [30]. Reproduced by permission of Rahman Amanullah and the AAS.

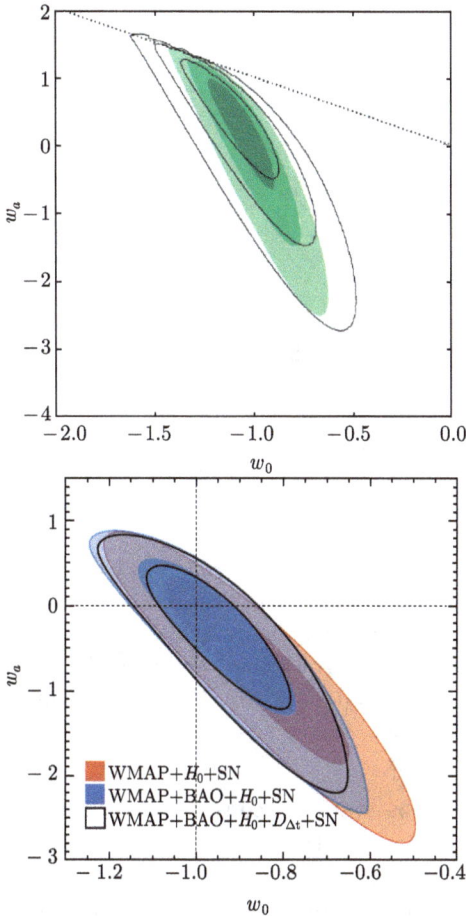

Figure 19.3   *Upper panel*: 68.3%, 95.4%, and 99.7% confidence regions of the (w_0, w_a) plane from Union2 SNIa sample combined with the constraints from BAO and CMB both with (solid contours) and without (shaded contours) systematic errors, for a flat universe. Points above the dotted line $(w_0 + w_a = 0)$ violate early matter domination and are implicitly disfavored in this analysis by the CMB and BAO data. From [30]. Reproduced by permission of Rahman Amanullah and the AAS. *Lower panel*: Joint constraints on the CPL model from the WMAP7 observations. The contours show 68.3% and 95.4% CL from WMAP+H_0+SNIa (red), WMAP+BAO+H_0+SNIa (blue) and WMAP+BAO+H_0+$D_{\Delta t}$+SNIa (black), for a flat universe. "$D_{\Delta t}$" denotes the time-delay distance to the lens system B1608+656 at $z = 0.63$ measured by [35]. From [29]. Reprinted by permission of Eiichiro Komatsu and the AAS.

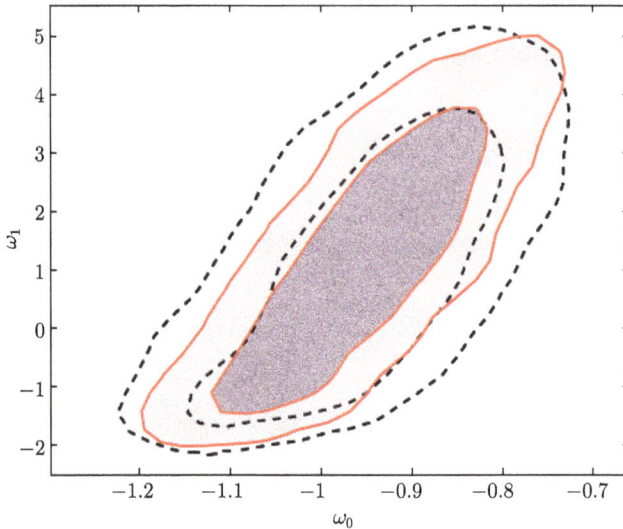

Figure 19.4   Joint two-dimensional marginalized constraint on the parameters w_0 and w_1 of the MZ parametrization. The contours show the 68% and 95% CL from Union2+WMAP+SDSS, for the cases without the systematic errors of SN (color shaded regions and red solid lines) and with the systematic errors of SN (unshaded regions and black dashed lines). Reprinted from [37], with permission of Xin Zhang and Elsevier.

we show the latest observational constraints on the MZ parametrization [37]. It has been proven that the MZ parametrizations is very successful in exploring the dynamical evolution of dark energy [36, 37], and has powerful prediction capability for the ultimate fate of the universe [38].

In addition, some other ansatzs have also been proposed. Using the principal component analysis, Linder and Huterer [39] argued that 2 parameters, involving a measure of the EOS value at some epoch (e.g. w_0) and a measure of the change in EOS (e.g. w'_{de}), are most realistic in projecting dark energy parameter constraints. Therefore, most ansatzs contain 2 parameters. In [40], Huterer and Turner proposed a linear parametrization $w_{\text{de}}(z) = w_0 + w_1 z$ to study the evolution of dark energy. This ansatz can fit the low redshifts data well, but its dark energy component grows increasingly unsuitable at redshifts $z > 1$. In [7] Jassal et $al.$ proposed a more general form of the CPL parametrization $w_{\text{de}}(z) = w_0 + w_a z/(1+z)^p$ and investigated the case of $p = 1, 2$. In [41], Efstathiou introduced another parametrization $w_{\text{de}}(z) = w_0 + w_1 \ln(1+z)$. It should be mentioned that the parametrizations listed above have difficulty to fit the rapidly varying dark energy models, and some other parametrizations [6, 42–44] have been proposed to fit a fast transition of $w_{de}(z)$. For more research works of various parametrization forms, see [2, 42, 45–48] and references therein.

19.2   Binned Parametrization

In addition to the specific ansatz, another popular approach is the binned parametrization. The binned parametrization was firstly proposed by Huterer and Starkman [49] based on the principal component analysis (PCA) [49, 50]. It is often used to measure the EOS w_{de} and the density ρ_{de} of dark energy. The key idea is dividing the redshift range into different bins and picking a simple local basis representation for $w_{de}(z)$ or $\rho_{de}(z)$. The simplest way is setting $w_{de}(z)$ or $\rho_{de}(z)$ as piecewise constant in redshift. For the case where w is piecewise constant in redshift, $f(z)$ can be written as [51]

$$f(z_{n-1} < z \leqslant z_n) = (1+z)^{3(1+w_n)} \prod_{i=0}^{n-1}(1+z_i)^{3(w_i-w_{i+1})}, \qquad (19.6)$$

where w_i is the EOS parameter in the i-th redshift bin defined by an upper boundary at z_i. This parametrization has been extensively studied [25, 52, 53]. For the case where dark energy density ρ_{de} is piecewise constant in redshift, $f(z)$ can be written as

$$f(z) = \begin{cases} 1, & 0 \leqslant z \leqslant z_1; \\ f_i, & z_{i-1} \leqslant z \leqslant z_i \ (2 \leqslant i \leqslant n). \end{cases} \qquad (19.7)$$

Here f_i is a piecewise constant, and from the relation $E(0) = 1$ one can easily obtain $f_1 = 1$. For same number of redshift bins, the number of free parameters of piecewise constant ρ_{de} parametrization is one fewer than that of piecewise constant w parametrization.

It should be mentioned that the optimal choice of redshift bins is still in debate. In [54], Riess et $al.$ proposed a uniform, unbiased binning method, in which the number of SNIa in each bin times the width of each bin is a constant (i.e. $n\Delta z = $ const.) Using $n\Delta z = 40$, 20, and 15, respectively, they derived 3, 4, or 5 bins' independent measurements of $H(z)$ and $\dot{a}$ from the gold sample [54] (See Figure 19.5). This binning method has drawn a lot of attention. For examples, by setting $n\Delta z \sim 30$ and using the piecewise

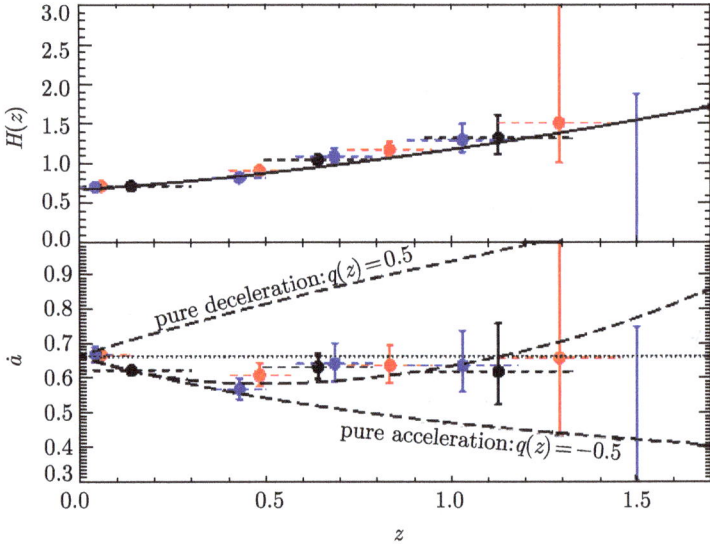

Figure 19.5   Uncorrelated estimates of the expansion history. Using $n\Delta z = 40$, 20, and 15, respectively, 3, 4, or 5 bins' independent measurements of $H(z)$ from the gold sample are plotted in the top panel. The solid black line in this plane denotes the prediction of the ΛCDM model with $\Omega_{m0} = 0.29$. The bottom panel shows the independent measurements of the kinematic quantity $\dot{a}$ versus redshift. In this plane a positive or negative sign of the slope of the data indicates deceleration or acceleration of the expansion, respectively. From [54]. Reproduced by permission of Adam Riess and the AAS.

constant w parametrization, Gong *et al.* explored the Constitution dataset in [55], and analyzed the Union2 dataset in [44].

In [56], Wang argued that one should choose a constant Δz for redshift slices. This is because for a galaxy redshift survey, the observables are H and $1/D_A$ (length scales extracted from data analysis). Since these scales are assumed to be constant in each redshift slice, the redshift slices should be chosen such that the variations of H and $1/D_A$ in each redshift slice remain roughly constant with z. This binning method has also been adopted by some experimental groups. For example, in [30], the SCP SNIa group explored the Union2 dataset by using this constant binning method. They found that the current small sample of SNIa cannot constrain the existence of dark energy above redshift 1 (see Figure 19.6).

In [57, 58], we presented a new binned parametrization method. Instead of choosing the discontinuity points z_i by hand, one can treat z_i as models

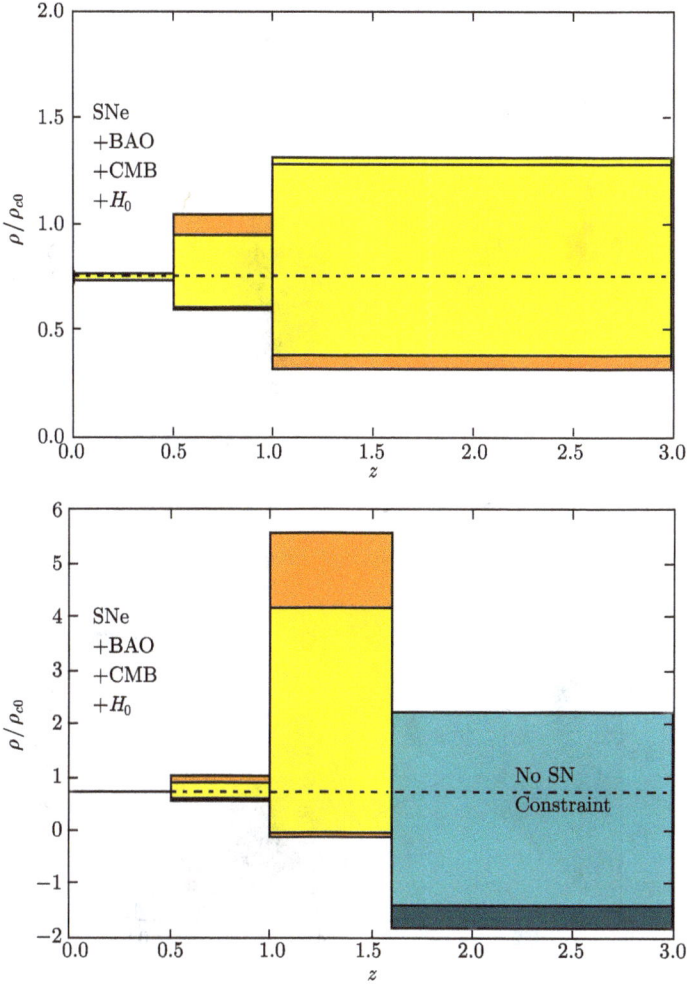

Figure 19.6   Constrains on the dark energy density from a joint data set of SN, BAO, CMB, and H_0. The upper panel is the case for three bins, the lower panel is the case for four bins. From [30]. Reproduced by permission of Rahman Amanullah and the AAS.

parameters and let them run freely in the redshift region of SNIa samples. Using the piecewise constant w and the piecewise constant ρ_{de} parametrization, respectively, the Constitution SNIa dataset has been explored [57]. In addition, utilizing the piecewise constant ρ_{de} parametrization, the Union2 SNIa dataset has also been analyzed [58] (the corresponding results are

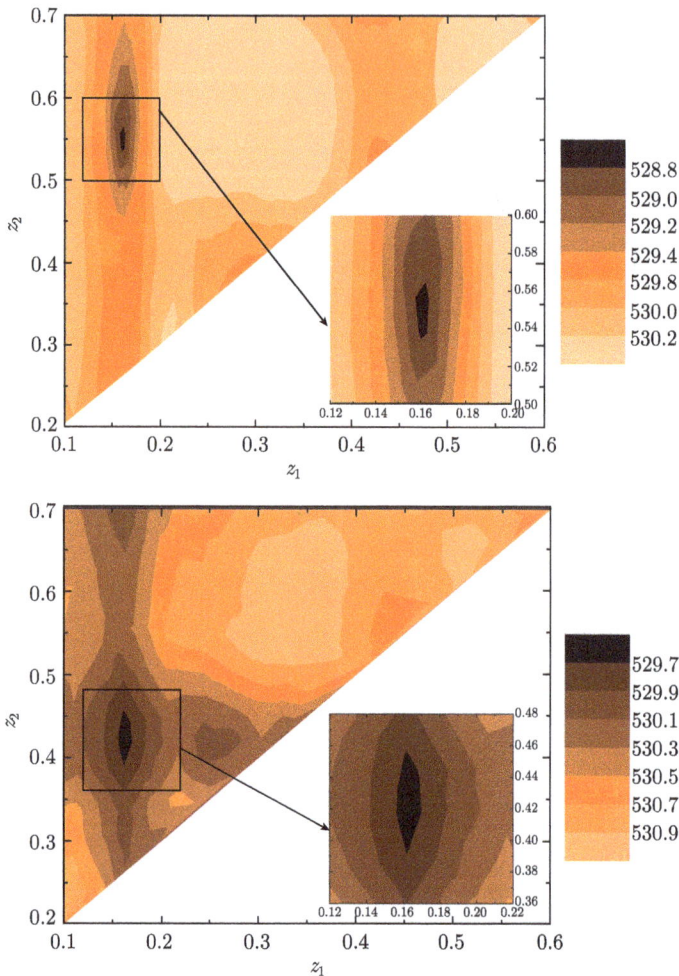

Figure 19.7   The relationship between the $\chi^2_{\min}$ and the discontinuity points of redshift (z_1 and z_2) for the 3 bins piecewise constant ρ_{de} parametrization. The upper panel is plotted by using the Union2 SNIa sample alone, and the lower panel is plotted by using the combined SNIa+CMB+BAO data. The x-axis represents the redshift of the first discontinuity point z_1, while the y-axis denotes the redshift of the second discontinuity point z_2. Notice that the light-colored region corresponds to a big χ^2, and the dark-colored region corresponds to a small χ^2. Since $z_1 \leqslant z_2$ must be satisfied, the bottom-right region of the figure is always blank. From [58].

given in Figure 19.7). These works show that the Constitution dataset favors a dynamical dark energy, while the Union2 dataset is still consistent with a cosmological constant. Comparing with those two binning methods listed above, the advantage of this binning method is that it can achieve much smaller $\chi^2_{\min}$.[1]

Besides the piecewise constant parametrization, some other local basis representations for $w_{\mathrm{de}}(z)$ or $\rho_{\mathrm{de}}(z)$ are also proposed, such as wavelet [60] and numerical derivatives [1, 61].

19.3 Polynomial Fitting

The third approach is the polynomial fitting method. The key idea is treating the dark energy density function $f(z)$ as a free function of redshift and representing it by using the polynomial. Compared with the binned parametrization, the advantage of the polynomial fitting parametrization is that the dark energy density function $f(z)$ can be reconstructed as a continuous function in the redshift range covered by the observational data.

A simple polynomial fit to $f(z)$ was proposed by Alam *et al.* [62], which is a truncated Taylor expansion

$$f(z) = A_0 + A_1(1 + z) + A_2(1 + z)^2. \tag{19.8}$$

This ansatz has only three free parameters $(\Omega_{\mathrm{m}0}, A_1, A_2)$ since $A_0 + A_1 + A_2 = 1 - \Omega_{\mathrm{m}0}$ for a flat universe. By using this ansatz, Alam *et al.* argued that the Tonry/Barris SNIa sample [63, 64] appears to favor dark energy which evolves in time [65] (see Figure 19.8). This conclusion will be modified if the effect of the CMB/LSS observations is taken into account [66]. It should be mentioned that there was a debate about the reliability of this ansatz [67, 68].

Another interesting polynomial fit is the polynomial interpolation, which was proposed by Wang [11–16, 69]. It chooses different redshift points $z_i = i * z_{\max}/n (i = 1, 2, \ldots, n)$, and interpolate $f(z)$ by using its own values at these redshift points. This yields

$$f(z) = \sum_{i=1}^{n} f_i \frac{(z - z_1) \cdots (z - z_{i-1})(z - z_{i+1}) \cdots (z - z_n)}{(z_i - z_1) \cdots (z_i - z_{i-1})(z_i - z_{i+1}) \cdots (z_i - z_n)}. \tag{19.9}$$

Here $f_i = f(z_i)$ and $z_n = z_{\max}$. Based on the relation $f(0) = 1$, one parameter can be fixed directly, and only $n - 1$ model parameters need to

---

[1] A simple comparison of these three binning methods can be seen in [59].

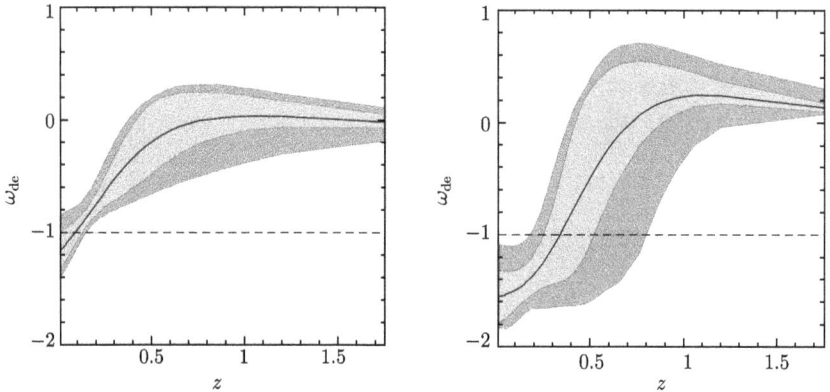

Figure 19.8   The evolution of $w_{de}(z)$ with redshift for different values of Ω_{m0}. The left panel corresponds to the case of $\Omega_{m0} = 0.2$, and the right panel corresponds to the case of $\Omega_{m0} = 0.4$, In each panel, the thick solid line shows the best-fit, the light grey contour represents the 1σ confidence level, and the dark grey contour represents the 2σ confidence level around the best-fit. Copyright (2008) Wiley. Used with permission from Ujjaini Alam [65].

be determined by the data. In [12], Wang and Tegmark made an accurate measurement of the dark energy density function $f(z)$ by using the spectacular of the high redshift supernova observations from the HST/GOODS program and previous supernova. In [13], Wang and Freese demonstrated that $\rho_{de}(z)$ can be constrained more tightly than $w_{de}(z)$ given the same observational data by using the Tonry/Barris SNIa sample [63, 64]. In [16], by utilizing the nearby+SDSS+ESSENCE+SNLS+HST set of 288 SNIa and the Constitution set of 397 SNIa, Wang showed that flux-averaging of SNIa can be used to test the presence of unknown systematic uncertainties, and yield more robust distance measurements from SNIa (see Figure 19.9 for details). The latest Union2 set of 557 SNIa has also been explored by using this polynomial interpolation method [58].

19.4   Gaussian Process Modeling

The fourth approach is the Gaussian Process (GP) modeling, which was proposed by Holsclaw *et al.* [70, 71]. GP is a stochastic process, which is indexed by z. The defining property of a GP is that the vector that corresponds to the process at any finite collection of points follows a multivariate Gaussian distribution [72]. GPs are elements of an infinite dimensional

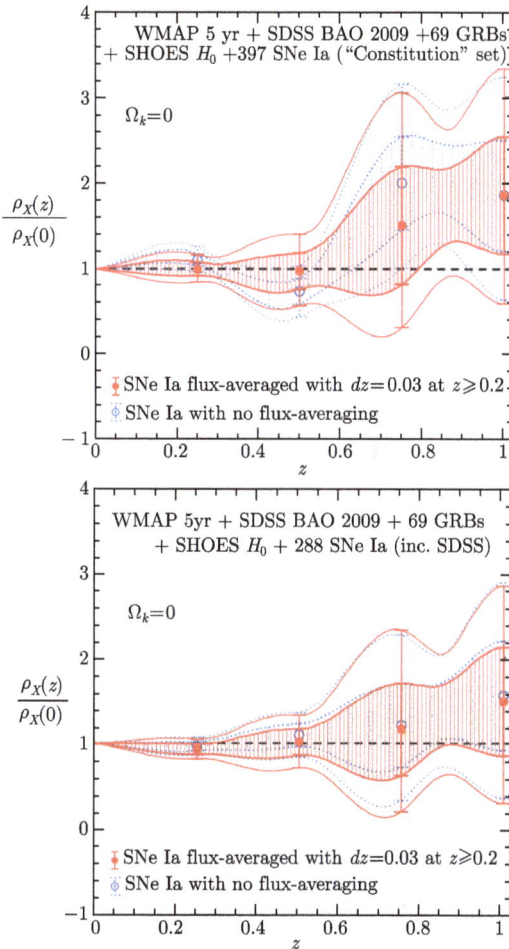

Figure 19.9   Dark energy density function $f(z) \equiv \rho_X(z)/\rho_X(0)$ measured from combining SNIa+CMB+BAO+GRB+H_0 data. The 68% (shaded) and 95% confidence level regions are shown. A flat universe is assumed. The upper panel is plotted by using the Constitution set of 397 SNIa, while the lower panel is plotted by using the nearby+SDSS+ESSENCE+SNLS+HST data set of 288 SNIa. The latter dataset gives much more stringent constraints, and gives measurements that are closer to a cosmological constant. Besides, flux-averaging has larger impact on the results from using the Constitution set, and brings the measurements closer to that predicted by a cosmological constant. From [16]. Reprinted figures with kind permission of Yun Wang and the APS. Copyright (2009) by the American Physical Society.

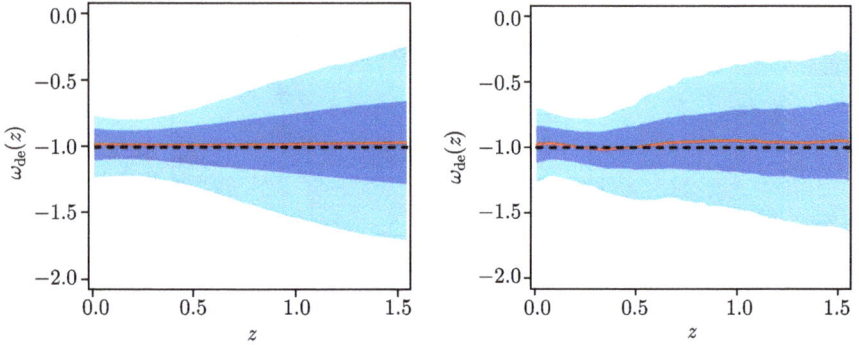

Figure 19.10   Reconstruction of $w_{de}(z)$ based on GP modeling combined with MCMC. The left panel uses a Gaussian covariance function ($\alpha \simeq 2$), while the right panel uses an exponential covariance function ($\alpha = 1$). These two results are very similar, both are very close and in agreement with a cosmological constant (black dashed line). The dark blue shaded region indicates the 68% CL, while the light blue region extends it to 95% CL. From [70]. Reprinted figures with kind permission of Ujjaini Alam, Katrin Heitmann and the APS. Copyright (2010) by the American Physical Society.

space, and can be used as the basis for a nonparametric reconstruction method. They are characterized by a mean and a covariance function, defined by a small number of hyperparameters [72, 73].

Based on the definition of a GP, one can assume that, for any collection $z_1, \ldots, z_n$, $w_{de}(z_1), \ldots, w_{de}(z_n)$ follow a multivariate Gaussian distribution with a constant negative mean and exponential covariance function written as

$$K(z, z') = \kappa^2 x^{|z-z'|^\alpha}. \tag{19.10}$$

The hyperparameters $x \in (0, 1)$ and κ, and the parameters defining the likelihood, are determined by the data. The value of $\alpha \in (0, 2]$ influences the smoothness of the GP realizations: for $\alpha = 2$, the realizations are smooth with infinitely many derivatives, while $\alpha = 1$ leads to rougher realizations suited to modeling continuous non-(mean-squared)-differentiable functions. Moreover, one can set up the following GP for w_{de}:

$$w_{de}(u) \sim GP(-1, K(u, u')). \tag{19.11}$$

Making use of the Eq. (19.11), one can take advantage of the particular integral structure of luminosity distance $d_L(z)$ expressed by $w_{de}(z)$ (see [70, 71] for details).

This new, nonparametric reconstruction method has the following advantages: it avoids artificial biases due to restricted parametric assumptions for $w_{de}(z)$, it does not lose information about the data by smoothing it, and it does not introduce arbitrariness (and lack of error control) in reconstruction by representing the data using a certain number of bins, or cutting off information by using a restricted set of basis functions to represent the data [70, 71]. In [70], using this reconstruction method, Holsclaw *et al.* reconstructed $w_{de}(z)$ utilizing the Constitution dataset [74]. The obtained results are consistent with the cosmological constant, with no evidence for a systematic mean evolution in w_{de} with redshift (see Figure 19.10).

References

[1] R. A. Daly and S. G. Djorgovski, ApJ. **597** (2003) 9.
[2] J. Weller and A. Albrecht, Phys. Rev. D **65** (2002) 103512.
[3] Y. G. Gong and A. Wang, Phys. Rev. D **75** (2007) 043520.
[4] D. Huterer and M. S. Turner, Phys. Rev. D **64** (2001) 123527.
[5] I. Maor, R. Brustein, and P. J. Steinhardt, Phys. Rev. Lett. **86** (2001) 6; I. Maor *et al.*, Phys. Rev. D **65** (2002) 123003.
[6] C. Wetterich, Phys. Lett. B **594** (2004) 17.
[7] H. K. Jassal, J. S. Bagla, and T. Padmanabhan, MNRAS **356** (2005) L11.
[8] A. Shafieloo, V. Sahni, and A. A. Starobinsky, Phys. Rev. D, **80** (2009) 101301.
[9] G. Efstathiou, MNRAS **310** (1999) 842; P. Astier, Phys. Lett. B **500** (2001) 8; J. Weller and A. Albrecht, Phys. Rev. Lett. **86** (2001) 1939; M. Tegmark, Phys. Rev. D **66** (2002) 103507; D. N. Spergel and G. D. Starkman, arXiv:astro-ph/0204089; S. Corasaniti and E. J. Copeland, Phys. Rev. D **67** (2003) 063521; Y. G. Gong and A. Wang, Phys. Rev. D **73** (2006) 083506; C. Clarkson and C. Zunckel, arXiv:1002.5004.
[10] M. S. Turner and M. J. White, Phys. Rev. D **56** (1997) 4439.
[11] Y. Wang and P. Garnavich, ApJ. **552** (2001) 445.
[12] Y. Wang and M. Tegmark, Phys. Rev. Lett. **92** (2004) 241302.
[13] Y. Wang and K. Freese, Phys. Lett. B **632** (2006) 449.
[14] Y. Wang, Phys. Rev. D **77** (2008) 123525.
[15] Y. Wang and P. Mukherjee, Phys. Rev. D **76** (2007) 103533.
[16] Y. Wang, Phys. Rev. D **80** (2009) 123525.
[17] E. V. Linder, Phys. Rev. D **70** (2004) 061302.
[18] F. Y. Wang, Z. G. Dai, and S. Qi, Astron. Astrophys. **507** (2009) 53.
[19] Z. X. Li, P. X. Wu, and H. W. Yu, Phys. Lett. B **695** (2011) 1; Z. X. Li, P. X. Wu, and H. W. Yu, arXiv:1011.2036.
[20] J. C. Carvalho and J. S. Alcaniz, arXiv:1102.5319.
[21] B. Santos, J. C. Carvalho, and J. S. Alcaniz, arXiv:1009.2733.
[22] V. Sahni *et al.*, JETP Lett. **77** (2003) 201.

[23] M. Visser, Class. Quant. Grav. **21** (2004) 2603.
[24] D. Rapetti *et al.*, MNRAS **375** (2007) 1510.
[25] C. Zunckel and C. Clarkson, Phys. Rev. Lett. **101** (2008) 181301.
[26] V. Sahni, A. Shafieloo, and A. A. Starobinsky, Phys. Rev. D **78** (2008) 103502.
[27] A. V. Pan and U. Alam, arXiv:1012.1591.
[28] E. Komatsu *et al.*, ApJS. **180** (2009) 330.
[29] E. Komatsu *et al.*, arXiv:1001.4538.
[30] R. Amanullah *et al.*, ApJ. **716** (2010) 712.
[31] M. Chevallier and D. Polarski, Int. J. Mod. Phys. D **10** (2001) 213.
[32] E. V. Linder, Phys. Rev. Lett. **90** (2003) 091301.
[33] E. V. Linder, Phys. Rev. D **70** (2004) 023511.
[34] R. Lazkoz, S. Nesseris, and L. Perivolaropoulos, JCAP **0807** (2008) 012; S. Basilakos, S. Nesseris, and L. Perivolaropoulos, MNRAS **387** (2008) 1126.
[35] S. H. Suyu *et al.*, ApJ. **711** (2010) 201.
[36] J. Z. Ma and X. Zhang, Phys. Lett. B **699** (2011) 233.
[37] H. Li and X. Zhang, Phys. Lett. B **703** (2011) 119.
[38] X. D. Li *et al.*, In preparation.
[39] E. V. Linder and D. Huterer, Phys. Rev. D **72** (2005) 043509.
[40] D. Huterer and M. S. Turner, Phys. Rev. D **60** (1999) 081301.
[41] G. Efstathiou, MNRAS **342** (2000) 810.
[42] U. Seljak *et al.*, Phys. Rev. D **71** (2005) 103515.
[43] A. Upadhye, M. Ishak, and P. J. Steinhardt, Phys. Rev. D **72** (2005) 063501; K. Ichikawa and T. Takahashi, JCAP **0702** (2007) 001; K. Ichikawa and T. Takahashi, JCAP **0804** (2008) 027; Y. G. Gong, B. Wang, and R. G. Cai, JCAP **1004** (2010) 019; N. N. Pan *et al.*, Class. Quant. Grav. **27** (2010) 155015.
[44] Y. G. Gong *et al.*, arXiv:1008.5010.
[45] R. Lazkoz, S. Nesseris, and L. Perivolaropoulos, JCAP **0511** (2005) 010.
[46] S. Nesseris and L. Perivolaropoulos, Phys. Rev. D **70** (2004) 043531.
[47] B. A. Bassett *et al.*, MNRAS **336** (2002) 1217; B. A. Bassett, P. S. Corasaniti and M. Kunz, ApJ. **617** (2004) L1; B. A. Bassett *et al.*, JCAP **0807** (2008) 007.
[48] R. Silva *et al.*, arXiv:1104.1628.
[49] D. Huterer and G. Starkman, Phys. Rev. Lett. **90** (2003) 031301.
[50] D. Huterer and A. Cooray, Phys. Rev. D **71** (2005) 023506.
[51] S. Sullivan, A. Cooray, and D. E. Holz, JCAP **09** (2007) 004.
[52] S. Qi, F. Y. Wang, and T. Lu, Astron. Astrophys. **483** (2008) 49.
[53] X. D. Li *et al.*, JCAP **07** (2011) 011.
[54] A. G. Riess *et al.*, ApJ. **659** (2007) 98.
[55] Y. G. Gong *et al.*, JCAP **01** (2010) 019.
[56] Y. Wang, Mod. Phys. Lett. A **25** (2010) 3093.
[57] Q. G. Huang *et al.*, Phys. Rev. D **80** (2009) 083515.
[58] S. Wang, X. D. Li, and M. Li, Phys. Rev. D **83** (2011) 023010.
[59] X. D. Li *et al.*, JCAP **07** (2011) 011.

[60] A. Hojjati, L. Pogosian, and G. B Zhao, JCAP **04** (2010) 1007.
[61] A. Shafieloo *et al.*, MNRAS **366** (2006) 1081.
[62] U. Alam *et al.*, MNRAS **344** (2003) 1057.
[63] J. L. Tonry *et al.*, ApJ. **594** (2003) 1.
[64] B. Barris *et al.*, ApJ. **602** (2004) 571.
[65] U. Alam *et al.*, MNRAS **354** (2004) 275.
[66] U. Alam, V. Sahni, and A. A. Starobinsky, JCAP **06** (2004) 008.
[67] J. Jonsson *et al.*, JCAP **0409** (2004) 007.
[68] U. Alam *et al.*, arXiv:astro-ph/0406672.
[69] Y. Wang and P. Mukherjee, ApJ. **606** (2004) 654; Y. Wang and M. Tegmark, Phys. Rev. D **71** (2005) 103513.
[70] T. Holsclaw *et al.*, Phys. Rev. Lett. **105** (2010) 241302.
[71] T. Holsclaw *et al.*, Phys. Rev. D **82** (2010) 103502.
[72] S. Banerjee, B. P. Carlin, and A. E. Gelfand, *Hierarchical Modeling and Analysis for Spatial Data* (Chapman and Hall New York: 2004); C. E. Rasmussen and K. I. Williams, *Gaussian Processes for Machine Learning* (MIT Press, Boston, 2006).
[73] S. Habib *et al.*, Phys. Rev. D **76** (2007) 083503.
[74] M. Hicken *et al.*, ApJ. **700** (2009) 1097; M. Hicken *et al.*, ApJ. **700** (2009) 331.

Index

www.ingramcontent.com/pod-product-compliance
Lightning Source LLC
Chambersburg PA
CBHW050550190326
41458CB00007B/1992